FE/EIT EXAM PR

CIVIL ENGINEERING

Fourth Edition

Braja M. Das, PhD, PE, California State University, Sacramento (Retired)

Indranil Goswami, PhD, PE, Morgan State University

Bruce E. Larock, PhD, PE, University of California, Davis (Emeritus)

Thomas Nelson, PhD, PE, University of Wisconsin, Platteville

Robert W. Stokes, PhD, Kansas State University

Michael Taylor, PhD, University of California, Davis (Emeritus)

Alan Williams, PhD, SE, California Department of Transportation

Kenneth J. Williamson, PhD, PE, Oregon State University

KAPLAN) AEC EDUCATION

This publication is designed to provide accurate and authoritative information in regard to the subject matter covered. It is sold with the understanding that the publisher is not engaged in rendering legal, accounting, or other professional service. If legal advice or other expert assistance is required, the services of a competent professional should be sought.

President: Mehul Patel
Vice President & General Manager: David Dufresne
Vice President of Product Development & Publishing: Evan M. Butterfield
Editorial Project Manager: Laurie McGuire
Director of Production: Daniel Frey
Senior Managing Editor, Production: Jack Kiburz
Creative Director: Lucy Jenkins
Production Artist: International Typesetting and Composition
Assistant Product Manager: Erica Smith

© 2008 by Dearborn Financial Publishing, Inc.®

Published by Kaplan AEC Education

30 South Wacker Drive, Suite 2500
Chicago, IL 60606-7481
(312) 836-4400
www.kaplanAECengineering.com

Printed in the United States of America

08 09 10   9 8 7 6 5 4 3 2 1

ISBN-13: 978-1-4277-6127-9
ISBN-10:   1-4277-6127-2

CONTENTS

CHAPTER 1 **Geotechnical** 1
PARTICLE SIZE DISTRIBUTION 1
WEIGHT-VOLUME RELATIONSHIPS 2
RELATIVE DENSITY 3
CONSISTENCY OF CLAYEY SOILS 3
PERMEABILITY 4
FLOW NETS 4
CONSOLIDATION 5
SHEAR STRENGTH 6
PROBLEMS 7
SOLUTIONS 8

CHAPTER 2 **Structural Analysis** 11
NEWTON'S LAWS 11
FREE BODY DIAGRAMS 13
TRUSSES AND FRAMES 15
PROBLEMS 19
SOLUTIONS 21

CHAPTER 3 **Hydraulics and Hydro Systems** 23
MANNING EQUATION 23
HAZEN-WILLIAMS EQUATION 25
REFERENCES 25

CHAPTER 4 **Structural Steel and Reinforced Concrete Design** 27
ELASTIC DESIGN OF STEEL BEAMS 27
COMPRESSION MEMBERS 31
TENSILE STRESS 34
STRENGTH DESIGN PRINCIPLES FOR REINFORCED CONCRETE MEMBERS 35

REFERENCES 44
PROBLEMS 45
SOLUTIONS 45

CHAPTER 5 — Wastewater and Solid Waste Treatment 47
WASTEWATER FLOWS 47
SEWER DESIGN 47
WASTEWATER CHARACTERISTICS 48
WASTEWATER TREATMENT 49
PROBLEMS 58
SOLUTIONS 59

CHAPTER 6 — Transportation Engineering 63
HIGHWAY CURVES 63
SIGHT DISTANCE 70
TRAFFIC CHARACTERISTICS 73
EARTHWORK 74
REFERENCES 76
PROBLEMS 78
SOLUTIONS 79

CHAPTER 7 — Water Purification and Treatment 83
WATER DISTRIBUTION 83
WATER QUALITY 85
WATER TREATMENT 91
PROBLEMS 95
SOLUTIONS 97

CHAPTER 8 — Surveying 101
GLOSSARY OF SURVEYING TERMS 102
BASIC TRIGONOMETRY 103
TYPES OF SURVEYS 103
COORDINATE SYSTEMS 103
STATIONING 104

CHAINING TECHNIQUES 104
DIFFERENTIAL LEVELING 105
ANGLES AND DISTANCES 106
TRAVERSE CLOSURE 109
AREA OF A TRAVERSE 111
AREA UNDER AN IRREGULAR CURVE 113
PROBLEMS 116
SOLUTIONS 118

CHAPTER 9 Construction Management 121

PROCUREMENT METHODS 122
CONTRACT TYPES 124
CONTRACTS AND CONTRACT LAW 124
CONSTRUCTION ESTIMATING 125
PRODUCTIVITY 129
PROJECT SCHEDULING 129
PROBLEMS 136
SOLUTIONS 137

APPENDIX Afternoon Sample Examination 139

Index 167

Introduction

OUTLINE

HOW TO USE THIS BOOK vii

BECOMING A PROFESSIONAL ENGINEER viii
Education ■ Fundamentals of Engineering/Engineer-in-Training Examination ■ Experience ■ Professional Engineer Examination

FUNDAMENTALS OF ENGINEERING/ENGINEER-IN-TRAINING EXAMINATION viii
Examination Development ■ Examination Structure ■ Examination Dates ■ Examination Procedure ■ Examination-Taking Suggestions ■ License Review Books ■ Textbooks ■ Examination Day Preparations ■ Items to take to the Examination ■ Special Medical Condition ■ Examination Scoring and Results ■ Errata

HOW TO USE THIS BOOK

Civil Engineering FE/EIT Exam Preparation is designed to help you prepare for the Fundamentals of Engineering/Engineer-in-Training exam. The book covers the discipline-specific afternoon exam in civil engineering. For the morning exam, Kaplan AEC offers the comprehensive review book, *Fundamentals of Engineering FE/EIT Exam Preparation*.

This book covers the major topics on the afternoon exam in civil engineering, reviewing important terms, equations, concepts, analysis methods, and typical problems. After reviewing the topic, you can work the end-of-chapter problems to test your understanding. Complete solutions are provided so that you can check your work and further refine your solution methodology.

After reviewing individual topics, you should take the Sample Exam at the end of the book. To alleviate anxiety about the actual test, you should simulate the exam experience as closely as possible. Answer the 60 questions in the Sample Exam in an uninterrupted four-hour period, without looking back at any content in the rest of the book. You may wish to consult the *Fundamentals of Engineering Supplied-Reference Handbook*, which is the only reference you are allowed to use in the actual exam.

When you've completed the Sample Exam, check the provided solutions to determine your correct and incorrect answers. This should give you a good sense of topics you may want to spend more time reviewing. Complete solution methods are shown, so you can see how to adjust your approach to problems as needed.

The following sections provide you with additional details on the process of becoming a licensed professional engineer and on what to expect at the exam.

BECOMING A PROFESSIONAL ENGINEER

To achieve registration as a Professional Engineer, there are four distinct steps: (1) education, (2) the Fundamentals of Engineering/Engineer-in-Training (FE/EIT) exam, (3) professional experience, and (4) the professional engineer (PE) exam. These steps are described in the following sections.

Education

Generally, no college degree is required to be eligible to take the FE/EIT exam. The exact rules vary, but all states allow engineering students to take the FE/EIT exam before they graduate, usually in their senior year. Some states, in fact, have no education requirement at all. One merely need apply and pay the application fee. Perhaps the best time to take the exam is immediately following completion of related coursework. For most engineering students, this will be the end of the senior year.

Fundamentals of Engineering/ Engineer-in-Training Examination

This eight-hour, multiple-choice examination is known by a variety of names—Fundamentals of Engineering, Engineer-in-Training (EIT), and Intern Engineer—but no matter what it is called, the exam is the same in all states. It is prepared and graded by the National Council of Examiners for Engineering and Surveying (NCEES).

Experience

States that allow engineering seniors to take the FE/EIT exam have no experience requirement. These same states, however, generally will allow other applicants to substitute acceptable experience for coursework. Still other states may allow a candidate to take the FE/EIT exam without any education or experience requirements.

Typically, four years of acceptable experience is required before one can take the Professional Engineer exam, but the requirement may vary from state to state.

Professional Engineer Examination

The second national exam is called Principles and Practice of Engineering by NCEES, but many refer to it as the Professional Engineer exam or PE exam. All states, plus Guam, the District of Columbia, and Puerto Rico, use the same NCEES exam. Review materials for this exam are found in other engineering license review books.

FUNDAMENTALS OF ENGINEERING/ ENGINEER-IN-TRAINING EXAMINATION

Laws have been passed that regulate the practice of engineering in order to protect the public from incompetent practitioners. Beginning in 1907 the individual states began passing *title* acts regulating who could call themselves engineers and offer services to the public. As the laws were strengthened, the practice of engineering was limited to those who were registered engineers, or to those working under the supervision of a registered engineer. Originally the laws were limited to civil engineering, but over time they have evolved so that the titles, and sometimes the practice, of most branches of engineering are included.

There is no national licensure law; licensure is based on individual state laws and is administered by boards of registration in each state. You can find a list of contact information for and links to the various state boards of registration at the Kaplan AEC Web site: *www.kaplanaecengineering.com*. This list also shows the exam registration deadline for each state.

Examination Development

Initially, the states wrote their own examinations, but beginning in 1966 NCEES took over the task for some of the states. Now the NCEES exams are used by all states. Thus, it is easy for engineers who move from one state to another to achieve licensure in the new state. About 50,000 engineers take the FE/EIT exam annually. This represents about 65% of the engineers graduated in the United States each year.

The development of the FE/EIT exam is the responsibility of the NCEES Committee on Examination for Professional Engineers. The committee is composed of people from industry, consulting, and education, all of whom are subject-matter experts. The test is intended to evaluate an individual's understanding of mathematics, basic sciences, and engineering sciences obtained in an accredited bachelor's degree of engineering. Every five years or so, NCEES conducts an engineering task analysis survey. People in education are surveyed periodically to ensure the FE/EIT exam specifications reflect what is being taught.

The exam questions are prepared by NCEES committee members, subject matter experts, and other volunteers. All people participating must hold professional licensure. When the questions have been written, they are circulated for review in workshop meetings and by mail. You will see mostly metric units (SI) on the exam. Some problems are posed in U.S. customary units (USCS) because the topics typically are taught that way. All problems are four-way multiple-choice questions.

Examination Structure

The FE/EIT exam is divided into a morning four-hour section and an afternoon four-hour section. There are 120 questions in the morning section and 60 in the afternoon.

The morning exam covers the topics that make up roughly the first $2^1/_2$ years of a typical engineering undergraduate program. All examinees take the same morning exam.

Seven different exams are in the afternoon test booklet, one for each of the following six branches: civil, mechanical, electrical, chemical, industrial, environmental. A general exam is included for those examinees not covered by the six engineering branches. Each of the six branch exams consists of 60 problems covering coursework in the specific branch of engineering. The general exam, also 60 problems, has topics that are similar to the morning topics. If you are taking the FE/EIT exam as a graduation requirement, your school may compel you to take the afternoon exam that matches the engineering discipline in which you are obtaining your degree. Otherwise, you can choose the afternoon exam you wish to take.

There are two approaches to deciding. One approach is to take the general afternoon exam regardless of your engineering discipline. Because the topics covered are similar to those in the morning exam, this approach may streamline your review time and effort. If you are still in college or recently graduated, these general topics may be very fresh in your mind.

The second approach is to take the afternoon exam that matches the discipline in which you majored. Particularly if you have been out of college for several years practicing this discipline in your daily work, you will be very familiar and comfortable with the topics. This may be to your advantage during your review time and in the pressure of the exam itself.

At the beginning of the afternoon test period, examinees will mark the answer sheet as to which branch exam they are taking. You could quickly scan the test, judge the degree of difficulty of the general versus the branch exam, then choose the test to answer. We do not recommend this practice, as you would waste time in determining which test to write. Further, you could lose confidence during this indecisive period.

Table I.1 summarizes the major topics for the civil engineering afternoon exam, including the percentage of problems you can expect to see on each one.

Table I.1 FE/EIT Civil Engineering Exam

Topic	Percentage of Problems
Construction management	10
Environmental engineering	12
Hydraulics and hydrologic systems	12
Materials	8
Soil mechanics and foundations	15
Structural analysis	10
Structural design	10
Surveying	11
Transportation	12

Examination Dates

NCEES prepares FE/EIT exams for use on a Saturday in April and October each year. Some state boards administer the exam twice a year; others offer it only once a year. The scheduled exam dates for the next ten years can be found on the NCEES Web site, *www.ncees.org/exams/schedules/*.

Those wishing to take the exam must apply to their state board several months before the exam date.

Examination Procedure

Before the morning four-hour session begins, the proctors pass out exam booklets and a scoring sheet to each examinee. Space is provided on each page of the examination booklet for scratchwork. The scratchwork will *not* be considered in the scoring. Proctors will also provide each examinee with a mechanical pencil for use in recording answers; this is the only writing instrument allowed. Do not bring your own lead or eraser. If you need an additional pencil during the exam, a proctor will supply one.

The examination is closed book. You may not bring any reference materials with you to the exam. To replace your own materials, NCEES has prepared a *Fundamentals of Engineering (FE) Supplied-Reference Handbook.* The handbook contains engineering, scientific, and mathematical formulas and tables for use in the examination. Examinees will receive the handbook from their state registration board prior to the examination. The *FE Supplied-Reference Handbook* is also included in the exam materials distributed at the beginning of each four-hour exam period.

There are three versions (A, B, and C) of the exam. These have the major subjects presented in a different order to reduce the possibility of examinees copying from one another. The first subject on your exam, for example, might be fluid mechanics, while the exam of the person next to you may have electrical circuits as the first subject.

The afternoon session begins following a one-hour lunch break. The afternoon exam booklets will be distributed along with a scoring sheet. There will be 60 multiple choice questions, each of which carries twice the grading weight of the morning exam questions.

If you answer all questions more than 15 minutes early, you may turn in the exam materials and leave. If you finish in the last 15 minutes, however, you must remain to the end of the exam period to ensure a quiet environment for all those still working, and to ensure an orderly collection of materials.

Examination-Taking Suggestions

Those familiar with the psychology of examinations have several suggestions for examinees:

1. There are really two skills that examinees can develop and sharpen. One is the skill of illustrating one's knowledge. The other is the skill of familiarization with examination structure and procedure. The first can be enhanced by a systematic review of the subject matter. The second, exam-taking skills, can be improved by practice with sample problems—that is, problems that are presented in the exam format with similar content and level of difficulty.

2. Examinees should answer every problem, even if it is necessary to guess. There is no penalty for guessing. The best approach to guessing is to first eliminate the one or two obviously incorrect answers among the four alternatives. If this can be done, the chance of selecting a correct answer obviously improves from 1 in 4 to 1 in 2 or 3.

3. Plan ahead with a strategy and a time allocation. There are 120 morning problems in 12 subject areas. Compute how much time you will allow for each of the 12 subject areas. You might allocate a little less time per problem for the areas in which you are most proficient, leaving a little more time in subjects that are more difficult for you. Your time plan should include a reserve block for especially difficult problems, for checking your scoring sheet, and finally for making last-minute guesses on problems you did not work. A time plan gives you the confidence of being in control. Misallocation of time for the exam can be a serious mistake. Your strategy might also include time allotments for two passes through the exam—the first to work all problems for which answers

are obvious to you, the second to return to the more complex, time-consuming problems and the ones at which you might need to guess.

4. Read all four multiple-choice answers options before making a selection. All distractors (wrong answers) are designed to be plausible. Only one option will be the best answer.

5. Do not change an answer unless you are absolutely certain you have made a mistake. Your first reaction is likely to be correct.

6. If time permits, check your work.

7. Do not sit next to a friend, a window, or other potential distraction.

License Review Books

To prepare for the FE/EIT exam you need two or three review books.

1. *Fundamentals of Engineering FE/EIT Exam Preparation* to provide a review of the common four-hour morning examination.

2. An afternoon supplement book if you are going to take one of the six branch exams.

3. *Fundamentals of Engineering (FE) Supplied-Reference Handbook*. At some point this NCEES-prepared book will be provided to applicants by their State Registration Board. You may want to obtain a copy sooner so you will have ample time to study it before the exam. You must, however, pay close attention to the *FE Supplied-Reference Handbook* and the notation used in it, because it is the only book you will have at the exam.

Textbooks

If you still have your university textbooks, they can be useful in preparing for the exam, unless they are out of date. To a great extent the books will be like old friends with familiar notation. You probably need both textbooks and license review books for efficient study and review.

Examination Day Preparations

The exam day will be a stressful and tiring one. You should take steps to eliminate the possibility of unpleasant surprises. If at all possible, visit the examination site ahead of time. Try to determine such items as

1. How much time should I allow for travel to the exam on that day? Plan to arrive about 15 minutes early. That way you will have ample time, but not too much time. Arriving too early, and mingling with others who are also anxious, can increase your anxiety and nervousness.

2. Where will I park?

3. How does the exam site look? Will I have ample workspace? Will it be overly bright (sunglasses), or cold (sweater), or noisy (earplugs)? Would a cushion make the chair more comfortable?

4. Where are the drinking fountains and lavatory facilities?

5. What about food? Most states do not allow food in the test room (exceptions for ADA). Should I take something along for energy in the exam? A light bag lunch during the break makes sense.

Items to Take to the Examination

Although you may not bring books to the exam, you should bring the following:

- *Calculator*—Beginning with the April 2004 exam, NCEES has implemented a more stringent policy regarding permitted calculators. For a list of permitted models, see the NCEES Web site (*www.ncees.org*). You also need to determine whether your state permits pre-programmed calculators. Bring extra batteries for your calculator just in case, and many people feel that bringing a second calculator is also a very good idea.

- *Clock*—You must have a time plan and a clock or wristwatch.

- *Exam Assignment Paperwork*—Take along the letter assigning you to the exam at the specified location to prove that you are the registered person. Also bring something with your name and picture (driver's license or identification card).

- *Items Suggested by Your Advance Visit*—If you visit the exam site, it will probably suggest an item or two that you need to add to your list.

- *Clothes*—Plan to wear comfortable clothes. You probably will do better if you are slightly cool, so it is wise to wear layered clothing.

Special Medical Condition

If you have a medical situation that may require special accommodation, you need to notify the licensing board well in advance of exam day.

Examination Scoring and Results

The questions are machine-scored by scanning. The answer sheets are checked for errors by computer. Marking two answers to a question, for example, will be detected and no credit will be given.

Your state board will notify you whether you have passed or failed roughly three months after the exam. Candidates who do not pass the exam the first time may take it again. If you do not pass, you will receive a report listing the percentages of questions you answered correctly for each topic area. This information can help focus the review efforts of candidates who need to retake the exam.

The FE/EIT exam is challenging, but analysis of previous pass rates shows that the majority of candidates do pass it the first time. By reviewing appropriate concepts and practicing with exam-style problems, you can be in that majority. Good luck!

Errata

The authors and publisher of this book have been careful to avoid errors, employing technical reviewers, copyeditors, and proofreaders to ensure the material is as flawless as possible. Any known errata and corrections are posted on the product page at our Web site, *www.kaplanAECengineering.com.* If you believe you have discovered an inaccuracy, please notify the engineering editor at Kaplan AEC Education:

Engineeringpress@kaplan.com
Fax: 312-836-9958
Kaplan AEC Education
30 S. Wacker Drive, Suite 2500
Chicago, IL 60606

CHAPTER 1

Geotechnical

Braja M. Das

OUTLINE

PARTICLE SIZE DISTRIBUTION 1

WEIGHT-VOLUME RELATIONSHIPS 2

RELATIVE DENSITY 3

CONSISTENCY OF CLAYEY SOILS 3

PERMEABILITY 4

FLOW NETS 4

CONSOLIDATION 5

SHEAR STRENGTH 6

PROBLEMS 7

SOLUTIONS 8

PARTICLE SIZE DISTRIBUTION

The particle size distribution in a given soil is determined in the laboratory by sieve analysis and hydrometer analysis. For classification purposes, in coarse-grained soils the following two parameters can be obtained from a particle size distribution curve:

$$\text{Uniformity coefficient, } c_u = \frac{D_{60}}{D_{10}} \tag{1.1}$$

$$\text{Coefficient of gradation, } c_c = \frac{D_{30}^2}{D_{60} \times D_{10}} \tag{1.2}$$

where D_{10}, D_{30}, D_{60} = diameters through which, respectively, 10 percent, 30 percent, and 60 percent of the soil pass.

WEIGHT-VOLUME RELATIONSHIPS

Soils are three-phase systems containing soil solids, water, and air (Figure 1.1).

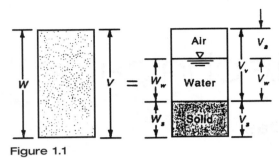

Figure 1.1

Referring to the figure,

$$W = W_s + W_w \tag{1.3}$$

$$V = V_s + V_v = V_s + V_w + V_a \tag{1.4}$$

where W = total weight of the soil specimen, W_s = weight of the solids, W_w = weight of water, V = total volume of the soil, V_s = volume of soil solids, V_v = volume of voids, V_w = volume of water, and V_a = volume of air.

The **volume relationships** can then be given as follows:

$$\text{Void ratio} = e = \frac{V_v}{V_s} \tag{1.5}$$

$$\text{Porosity} = n = \frac{V_v}{V} = \frac{e}{1+e} \tag{1.6}$$

$$\text{Degree of saturation} = S(\%) = \frac{V_w}{V_v} \times 100 = \frac{wG}{e} \times 100 \tag{1.7}$$

Similarly, the **weight relationships** are

$$\text{Water content (or moisture content)}(\%) = w = \frac{W_w}{W_s} \times 100 \tag{1.8}$$

$$\text{Moist unit weight} = \gamma = \frac{W}{V} = \frac{G\gamma_w(1+w)}{1+e} \tag{1.9}$$

$$\text{Dry unit weight} = \gamma_d = \frac{W_s}{V} = \frac{\gamma}{1+w} = \frac{G\gamma_w}{1+e} \tag{1.10}$$

$$\text{Unit weight of solids} = \gamma_s = \frac{W_s}{V_s} = G\gamma_w \tag{1.11}$$

In the preceding equations, γ_w = unit weight of water (62.4 lb/ft³ or 9.81 kN/m³) and G = specific gravity of soil solids, or

$$G = \frac{W_s}{V_s \gamma_w} \tag{1.12}$$

RELATIVE DENSITY

In granular soils, the degree of compaction is generally expressed by a nondimensional parameter called **relative density**, D_d, or

$$D_d = \frac{e_{max} - e}{e_{max} - e_{min}} \quad (1.13)$$

where e = actual void ratio in the field, e_{max} = void ratio in the loosest state, and e_{min} = void ratio in the densest state.

In terms of dry unit weight,

$$D_d = \frac{\dfrac{1}{\gamma_{min}} - \dfrac{1}{\gamma_d}}{\dfrac{1}{\gamma_{min}} - \dfrac{1}{\gamma_{max}}} \quad (1.14)$$

where γ_{min}, γ_{max} = minimum and maximum *dry* unit weights, respectively and γ_d = dry unit weight in the field.

CONSISTENCY OF CLAYEY SOILS

The moisture content in percent at which the cohesive soil will pass from a liquid state to a plastic state is called the **liquid limit**. Similarly, the moisture contents at which the soil changes from a plastic state to a semisolid state and from a semisolid state to a solid state are referred to as the **plastic limit** and the **shrinkage limit**, respectively. These limits are referred to as the **Atterberg limits** (see Figure 1.2).

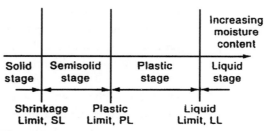

Figure 1.2

Liquid limit (LL) is the moisture content in percent at which the groove in the *Casagrande liquid limit device* closes for a distance of 0.5 in after 25 blows.

Plastic limit (PL) is the moisture content in percent at which the soil, when rolled into a thread of 1/8 in diameter, crumbles.

Plasticity index (PI) is defined as

$$PI = LL - PL \quad (1.15)$$

Shrinkage limit (SL) is the moisture content at which the volume of the soil mass no longer changes.

Shrinkage index (SI) is defined as

$$SI = PL - SL \quad (1.16)$$

PERMEABILITY

The rate of flow of water through a soil of gross cross-sectional area A can be given by the relationships shown in Figure 1.3, called Darcy's law,

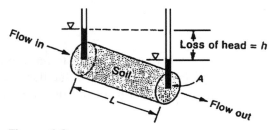

Figure 1.3

$$v = ki \qquad (1.17)$$

where v = discharge velocity, k = coefficient of permeability, i = hydraulic gradient = h/L (see Figure 1.3).

$$Q = vA = kiA \qquad (1.18)$$

where Q = flow through soil in unit time and A = area of cross section of the soil at a right angle to the direction of flow. Or

$$k = \frac{Q}{iA} \qquad (1.19)$$

FLOW NETS

In many cases, flow of water through soil varies in direction and in magnitude over the cross section. In those cases, calculation of rate of flow of water can be made by using a graph called a **flow net**. A flow net is a combination of a number of flow lines and equipotential lines. A flow line is one along which a water particle will travel from the upstream side to the downstream side. An equipotential line is one along which the potential head at all points is the same.

Figure 1.4 shows an example of a flow net in which water flows from the upstream to the downstream around a sheet pile. Note that in a flow net the flow lines and equipotential lines cross at *right angles*. Also, the flow elements constructed are *approximately square*. Referring to Figure 1.4, the flow line in unit

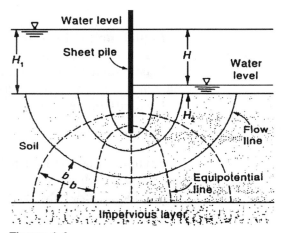

Figure 1.4

time (Q) per unit length normal to the cross section shown is

$$Q = k\frac{N_f}{N_d}H \qquad (1.20)$$

where N_f = number of flow channels, N_d = number of drops, and H = head difference between the upstream and downstream side. (Note that in Figure 1.4, $N_f = 4$ and $N_d = 6$.)

CONSOLIDATION

Consolidation settlement is the result of volume change in saturated clayey soils due to the expulsion of water occupied in the void spaces. In soft clays the major portion of the settlement of a foundation may be due to consolidation. Based on the theory of consolidation, a soil may be divided into two major categories: (a) normally consolidated and (b) overconsolidated. For *normally consolidated* clay, the *present effective overburden pressure* is the maximum to which the soil has been subjected in the recent geologic past.

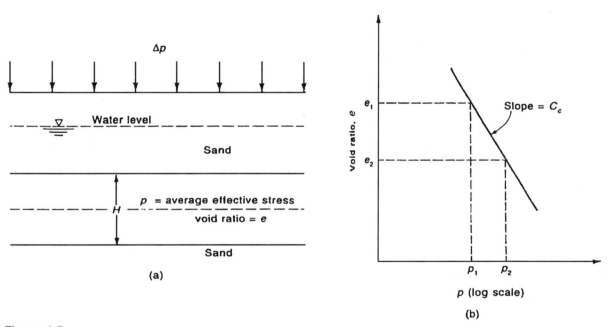

Figure 1.5

Figure 1.5(a) shows a normally consolidated clay deposit of thickness H. The consolidation settlement (ΔH) due to surcharge Δp can be determined as

$$\Delta H = \frac{C_c H}{1+e_i}\log\left(\frac{p_i + \Delta p}{p_i}\right) \qquad (1.21)$$

where e_i = initial void ratio, p_i = initial *average* effective pressure, and C_c = compression index.

The compression index can be determined from the laboratory as (see Figure 1.5(b))

$$C_c = \frac{e_1 - e_2}{\log\left(\frac{p_2}{p_1}\right)} \qquad (1.22)$$

SHEAR STRENGTH

The shear strength of soil (s), in general, is given by the **Mohr-Coulomb failure criteria**, or

$$s = c + \sigma' \tan \phi \tag{1.23}$$

where c = cohesion, σ' = effective normal stress, and ϕ = drained friction angle. For sands, $c = 0$, and the magnitude of ϕ varies with the relative density of compaction, size, and shape of the soil particles.

For normally consolidated clays, $c = 0$, so $s = \sigma' \tan \phi$. For overconsolidated clays, however, $c \neq 0$, so $s = c + \sigma' \tan \phi$. An important concept for the shear strength of cohesive soils is the so-called $\phi = 0$ concept. This is the condition in which drainage from the soil does not take place during loading. For such a case

$$s = c_u \tag{1.24}$$

where c_u is the undrained shear strength.

The *unconfined compression strength*, q_u, of a cohesive soil is

$$q_u = 2c_u \tag{1.25}$$

PROBLEMS

1.1 A moist soil specimen has a volume of 0.15 m^3 and weighs 2.83 kN. The water content is 12%, and the specific gravity of soil solids is 2.69. Determine
   a. Moist unit weight, γ
   b. Dry unit weight, γ_d
   c. Void ratio, e
   d. Degree of saturation, S

1.2 For a soil deposit in the field, the dry unit weight is 14.9 kN/m^3. From the laboratory, the following were determined: $G = 2.66$, $e_{max} = 0.89$, $e_{min} = 0.48$. Find the relative density in the field.

1.3 For a sandy soil, the maximum and minimum void ratios are 0.85 and 0.48, respectively. In the field, the relative density of compaction of the soil is 29.3 percent. Given $G = 2.65$, determine the moist unit weight of the soil at $w = 10\%$.

1.4 Refer to the flow net shown in Figure 1.4. Given $k = 0.03$ cm/min, $H_1 = 10$ m, and $H_2 = 1.8$ m, determine the seepage loss per day per meter under the sheet pile construction.

1.5 The results of a sieve analysis of a granular soil are as follows:

U.S. Sieve No.	Sieve Opening (mm)	Percent Retained on Each Sieve
4	4.75	0
10	2.00	20
40	0.425	20
60	0.25	30
100	0.15	20
200	0.075	5

Determine the uniformity coefficient and coefficient of gradation of the soil.

1.6 For a normally consolidated clay of 3.0 m thickness, the following are given:

$$\text{Average effective pressure} = 98 \text{ kN/m}^2$$
$$\text{Initial void ratio} = 1.1$$
$$\text{Average increase of pressure in the clay layer} = 42 \text{ kN/m}^2$$
$$\text{Compression index} = 0.27$$

Estimate the consolidation settlement.

1.7 An oedometer test in a normally consolidated clay gave the following results.

Average Effective Pressure (kN/m²)	Void Ratio
100	0.9
200	0.82

Calculate the compression index.

SOLUTIONS

1.1

a. $\gamma = \dfrac{W}{V} = \dfrac{2.83}{0.15} = 18.87 \text{ kN/m}^3$

b. $\gamma_d = \dfrac{\gamma}{1+w} = \dfrac{1887}{1+\left(\frac{12}{100}\right)} = 16.85 \text{ kN/m}^3$

c. $\gamma_d = \dfrac{G\gamma_w}{1+e}; \quad e = \dfrac{G\gamma_w}{\gamma_d} - 1 = \dfrac{(2.69)(9.81)}{16.85} - 1 = 0.566$

d. $S = \dfrac{wG}{e} \times 100 = \dfrac{(0.12)(2.69)}{0.566} \times 100 = 57.03\%$

1.2 In the field

$$\gamma_d = \dfrac{G\gamma_w}{1+e}; \quad e = \dfrac{G\gamma_w}{\gamma_d} - 1 = \dfrac{(2.66)(9.81)}{14.9} - 1 = 0.75$$

$$D_d = \dfrac{e_{max} - e}{e_{max} - e_{min}} = \dfrac{0.85 - 0.75}{0.89 - 0.48} = 34\%$$

1.3

$$D_d = 0.293 = \dfrac{e_{max} - e}{e_{max} - e_{min}} = \dfrac{0.85 - e}{0.85 - 0.45}; \quad e = 0.733$$

$$\gamma = \dfrac{G\gamma_w(1+w)}{1+e} = \dfrac{(2.65)(9.81)(1+0.1)}{1+0.733} = 16.5 \text{ kN/m}^3$$

1.4

$$Q = k\dfrac{N_f}{N_d}H = \dfrac{(0.03 \times 60 \times 24 \text{ cm/day})}{100}\left(\dfrac{4}{6}\right)(10-1.8) = 2.36 \text{ m}^3/\text{day/m}$$

1.5

Sieve Opening (mm)	Cumulative Percent Passing
4.75	100
2.0	80
0.425	60
0.25	30
0.15	10
0.075	5

So $D_{60} = 0.425$ mm; $D_{30} = 0.25$ mm; $D_{10} = 0.15$ mm

$$c_u = \frac{D_{60}}{D_{10}} = \frac{0.425}{0.15} = 2.83$$

$$c_c = \frac{D_{30}^2}{D_{60} \times D_{10}} = 0.98$$

1.6

$$\Delta H = \frac{C_c H}{1+e_i} \log\left(\frac{p_i + \Delta p}{p_i}\right) = \frac{(0.27)(3)}{1+1.1} \log\left(\frac{98+42}{98}\right) = 0.0597 \text{ m} = 59.7 \text{ mm}$$

1.7

$$C_c = \frac{e_1 - e_2}{\log\left(\frac{p_2}{p_1}\right)} = \frac{0.9 - 0.82}{\log\left(\frac{200}{100}\right)} = 0.266$$

CHAPTER 2

Structural Analysis

Michael Taylor

OUTLINE

NEWTON'S LAWS 11

FREE BODY DIAGRAMS 13

TRUSSES AND FRAMES 15
Trusses ■ Frames

PROBLEMS 19

SOLUTIONS 21

The outline notes that constitute this chapter are a summary of what can be found in all textbooks upon this subject. Since no one person's choice of summary can hope to match the needs of all students simultaneously, it is recommended that a suitable textbook be reread in conjunction with these notes and that personal notes be appended to the text in this chapter.

Determinate structural analysis, often termed **statics**, deals with structures that do not move. There are only two types of motion: translation and rotation. If the structure (and all parts of it) neither translates nor rotates, it is said to be in **static equilibrium**. (It is true, of course, that any material under stress will undergo some change of size and/or shape because of those stresses, but these movements are considered negligible in the present context.)

In statics, translation is caused by **forces**, and rotations are caused by **moments**. These are vector quantities. To define a vector requires three characteristics, usually (a) a line of action, (b) a direction, and (c) a magnitude. In some contexts, moments whose vectors are out of the plane are termed **torques** or **twists**. The student is expected to be familiar with simple vectors and their manipulation.

Actions is a general term that includes both forces and moments.

NEWTON'S LAWS

The study of statics is based upon two of Newton's three laws of motion. These are (1) a body will remain in its state of rest (relative to some chosen reference point) or of motion in a straight line, unless acted upon by a force, and (2) to every action there is an equal and opposite reaction. The third law (a body under the action of a single force will translate with acceleration along the line of the force and in the same direction as the force) is used in the study of *dynamics*.

Law 2 is often restated as "Forces (actions) can exist only in equal and opposite pairs." The combination of laws 1 and 2 thus requires that the net action on a body at rest be zero. This requirement, in turn, defines the two vector equations

$$\text{Net force} = 0$$
$$\text{Net moment} = 0$$

This is the central principle of statics.

More generally, determinate structural analysis can be defined as the process of calculating (for a body in static equilibrium)

(a) Any one, or all, of the actions acting upon the body

(b) The movement (deflection) of any point (in any direction) within the body

(c) The rotation of any line in the body

The process of structural analysis has just two steps. *First*, an idealized model of the structure (including the forces acting upon it) is agreed upon. In all but the most complex or large structures, it is adequate to use two-dimensional representations of real structures. When modeling the structure, it is also common practice to idealize the supports. Such idealizations as frictionless rollers, fully fixed supports, and frictionless pins are assumed to be familiar to the reader.

One other assumption is that structures are composed of rigid members. This does not mean that each member is infinitely rigid (i.e., does not change length or shape when loaded) but rather that these deformations are small relative to the original dimensions. Expressed another way, the deformations induced by loading do not change the geometry of the structure significantly. Luckily, this assumption is a good one for the vast majority of real engineering structures. The discerning reader may be about to object that many textbook examples (and reality) contain structures with ropes or wires, which are not stable if placed in compression. This objection is a good one but is countered by the observation (often implicit) that in all such cases the rope or wire must be in tension and hence cannot "buckle" or otherwise move.

Most engineers' drawings (and most examination problems) already include the simplifications just described.

The *second* step of the analysis process is the application of Newton's laws to the model.

The application of these equations of statics (often termed the equations of equilibrium) to a three-dimensional body involves just two vector equations (translation = 0 and rotation = 0). However, it is frequently more convenient to use the equivalent six independent scalar equations (three for translation and three for rotations). For two-dimensional models these reduce to three (two translations and one rotation). Thus, if a two-dimensional structural problem involves more than three unknowns, it is statically indeterminate and requires further knowledge for its solution. In what follows, the discussion is limited to two dimensions and stable static structures.

FREE BODY DIAGRAMS

The single most important concept in the *solution* of statics problems is that of the **free body diagram** (FBD). This concept is founded on the observation that if a (rigid) structure is in equilibrium, then every part of it must also be in equilibrium. This method has proven to minimize the likelihood of errors in solving structural problems.

In this approach an imaginary closed line is drawn through the structure. This line isolates a part of the structure, called the **free body**. The isolated part may be drawn by itself, but to maintain consistency with the original structure and loading, the engineer must examine the complete length of the closed line and insert all possible actions in their correct form and at their correct locations. Of course, some of these actions will be known and some unknown. All known actions must be included with their correct locations, magnitudes, and orientations. All unknown actions are given names, and their directions may be assumed (the mathematics will provide both the correct magnitude and the correct direction for each unknown).

It should be obvious that if the FBD is incorrect in *any way*, then the solution will be incorrect. A solution may be obtained, but if so, it is the solution to some *other* problem.

It is recommended that for each FBD used, a clear choice of axes be indicated and a positive direction also be chosen. These choices are, of course, arbitrary. It is then recommended that when applying the equations of equilibrium, all terms be written on one side of the equation and the result set equal to zero. The consistent use of this technique will reduce errors. By contrast, the practice, for example, of setting "all up forces equal to all down forces," which may seem attractive at first, is fraught with peril.

Facility in the analysis of structures is best obtained by solving multiple problems of as wide a variety as can be found. An excellent learning tool is to outline the steps of the solution without necessarily performing the mathematics. This permits a more rapid acquisition of the skills needed to understand how new problems are solved.

Note that in the following examples, units are not specified. The reader may select any units desired (e.g., metric versus "English" and small versus large).

In the following two problems the entire structure is used as the FBD.

Example 2.1

Find the reactions necessary to keep the structure shown in Exhibit 1 in equilibrium. Assume the axes, names, and directions shown in the figure.

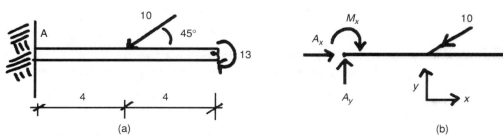

Exhibit 1 (a) Structure; (b) its FBD

Solution

Step 1. Sum forces in x direction:

$$A_x + (-).707 \times 10 = 0$$
$$A_x = +7.07 \text{ (assumed direction correct)}$$

Step 2. Sum forces in y direction:

$$A_y + (-).707 \times 10 = 0$$
$$A_y = +7.07 \text{ (assumed direction correct)}$$

Step 3. Sum moments about an axis through A (assume clockwise positive):

$$(+)M_x + (+)10 \times 4 \times .707 + (+)13 = 0$$
$$M_x = -41.3 \text{ (assumed direction was incorrect)}$$

This completes the solution requested.

Example 2.2

Find the reactions necessary to keep the structure shown in Exhibit 2 in equilibrium.

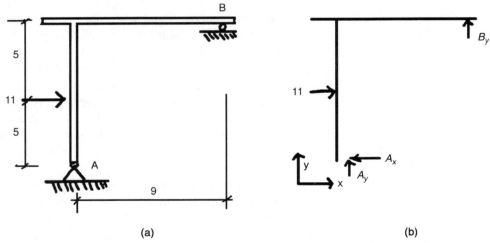

Exhibit 2 (a) Structure; (b) FBD of entire structure

Assume the names, axes, and directions shown.

Solution

Step 1. Summing forces in the positive x direction gives

$$+11 + (-)(A_x) = 0$$

Hence $(A_x) = +11$ (i.e., assumed direction is correct)

Step 2. Taking moments about an axis through A (with clockwise positive) gives

$$(-)B_y \times 9 + (+)11 \times 5 = 0$$

Hence

$$B_y = +\frac{55}{9} = +6.1$$

Step 3. Summing forces in the *y* direction gives

$$A_y + 6.1 = 0$$
$$A_y = -6.1 \text{ (i.e., assumed direction was incorrect)}$$

This completes the solution requested.

TRUSSES AND FRAMES

It should already be obvious to the reader that there is no single best way to solve structural analysis problems. It is true that a judicious choice of FBD and of the order in which equations are solved can shorten the calculation process, but all solution approaches will produce the correct answers if applied without error. However, it is convenient (but nothing more) to classify certain groups of structures, because they share characteristics and thus can be handled by a more "systematic" solution process. Two such groups are trusses and frames.

Trusses

Trusses must meet three criteria:

- All members (often termed **bars**) of the structure must be connected at only two points (these are called **joints**).
- All joints are frictionless pins.
- Loads are applied only at the joints.

The FBD of a bar of a truss can thus have only two forces acting upon it: one at each frictionless pin. Note that the bars need not be straight (although they almost always are). Application of the equations of statics shows that the line of action of both these forces must be directed along the line joining the two joints and that they must be equal in magnitude but oppositely directed. Hence a member of a truss is often described as a "two-force member." Only two possibilities then exist: Either the two forces are directed toward each other (in which case the bar is in compression) or they are directed away from each other (in which case the bar is in tension).

This feature of trusses simplifies their analysis, because if an FBD is drawn that cuts a member, it is known that only one action can exist at the cut, and it must be a force directed along the line that connects the joints (the sense and magnitude to be found as part of the solution).

There are two "classes" of analysis for trusses.

- **Method of Joints:** In this approach all FBDs are joints (with the exception that the entire body FBD is often used to find the reaction).

- **Method of Sections:** In this method the FBDs can include portions of the structure that are larger than a single joint. It should be noted that most designers use a combination of these methods when solving structures.

Example 2.3

Find the bar force in member CB in Exhibit 3(a) using the method of joints. Use an FBD of the entire structure, shown in Exhibit 3(a), to obtain reactions. Assume the axes, names, and directions shown.

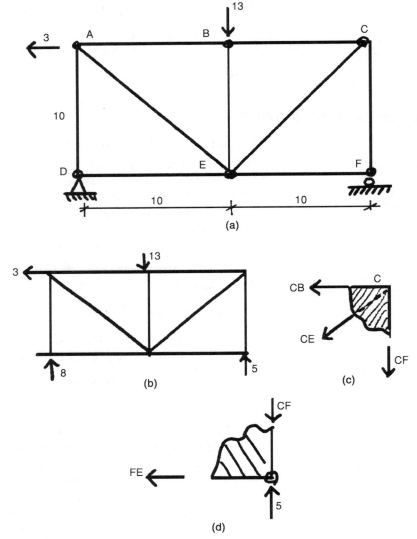

Exhibit 3

Solution

Solving by the methods shown above, the reader should obtain

$$D_y = 8.0 \text{ up}$$
$$F_y = 5.0 \text{ up}$$
$$D_x = 3 \text{ to right}$$

This completes the reactions, as shown in Exhibit 3(b).

Note that if we wish to attempt to find CB using an FBD of joint C, there will be three unknowns, which are not solvable; see Exhibit 3(c). We must find one of the other bar forces, CE or CF, in order to obtain a solution for CB. One possibility is to solve joint F first. This will give us CF and then permit a solution at joint C.

With Exhibit 3(d) and the assumed directions, joint F is solved to give CF = 5 (with bar in compression).

Now we move back to joint C in Exhibit 3(c) but with CF known in magnitude and direction—hence the arrow for CF in Exhibit 3(c) must be reversed. Taking moments about E with clockwise assumed positive,

$$CB = 5 \text{ and CB is in compression}$$

Example 2.4

Find the bar force in member DH in Exhibit 4(a) using the method of sections.

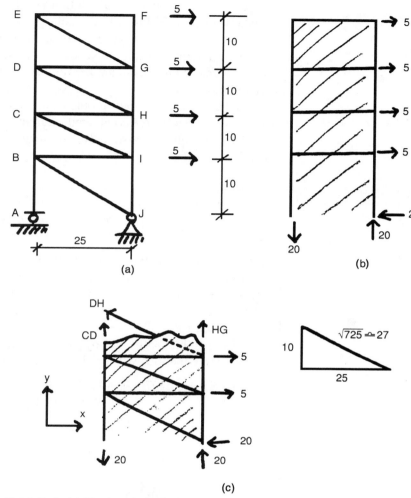

Exhibit 4 (a) Structure; (b) FBD entire structure; (c) FBD with horizontal cut through DC and GH

Solve to get the reactions shown in Exhibit 4(b).

Solution

Note that a horizontal cut can be used to give the FBD shown in Exhibit 4(c). Sum forces in x direction:

$$+5 + 5 - 20 - DH \times \frac{25}{27} = 0$$

$$DH = +10.8 \text{ (assumed direction correct, bar DH in tension)}$$

Frames

Frames are multimember structures that do not meet the criteria for trusses (although individual members within a structure may be two-force members and thus help to simplify their solution). Thus, when an FBD cuts a member, there are (in two dimensions) three possible actions across the cut, and these must be included in the FBD.

Frames are clearly more complicated to solve than trusses, and there are no simple rules or guidelines for solution other than (a) the general rule that several FBDs will be needed for a solution and (b) the designer must experiment with several possible FBDs to find a solution.

It is an excellent practice (though very infrequently followed) to outline the steps of a solution (i.e., the sequence of FBDs to be used and the identification of the unknowns that can be found at each step) prior to performing any calculations.

When relatively small problems are solved, it is always a good idea to make a quick "survey" of the completed solution to check for gross errors. The equations of statics can often be assessed approximately without the need for pencil and paper.

Example 2.5

Determine the force in member AC in Exhibit 5(a).

Solution

Note that this structure is externally indeterminate; that is, the reactions cannot be found by the equations of statics alone. We note, however that bar AC is a two-force member, so we can take BCD and FBD and solve for the force AC. (Note that we cannot solve the second FBD completely, but we can get the bar force asked for.)

From Exhibit 5(b) and by taking moments about B (note that neither of the translation equations will produce a numerical result), with clockwise positive:

$$(+)17 \times 17 + (-).707 \times CA \times 11 = 0$$
$$CA = 37.2 \text{ (assumed direction correct, CA in tension)}$$

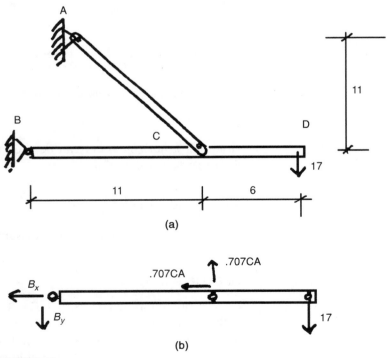

Exhibit 5

PROBLEMS

The reader is encouraged to attempt these problems alone prior to reviewing the sample solutions and to gain insight by comparing solutions where these differ.

In all cases the self-weight of the structure may be ignored unless instructions indicate otherwise.

Indicate the best answer from those offered (roundoff errors may lead to small discrepancies).

2.1 For the structure shown in Exhibit 2.1 the reactions are (in kN)

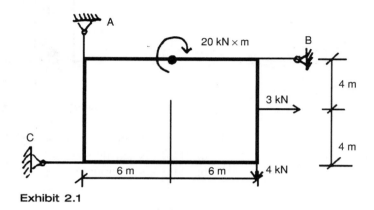

Exhibit 2.1

a. A = 4 kN down, B = 10 kN to left, C = 11.5 kN to right
b. A = 3 kN down, B = 10 kN to right, C = 10.0 kN to right
c. A = 4 kN up, B = 10 kN to left, C = 7.0 kN to right
d. A = 4 kN up, B = 8.5 kN to left, C = 11.5 kN to left

2.2 The truss shown in Exhibit 2.2 is used as a weighing device. If the gauge at A reads 9 kN, what is the weight of mass P?

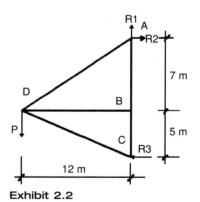

Exhibit 2.2

a. 9 kN
b. 12 kN
c. 4 kN
d. 8 kN

2.3 The force in member CB in Exhibit 2.3 is
  a. 4.65 kN tensile
  b. 4.65 kN compressive
  c. 47.5 kN tensile
  d. 8.00 kN compressive

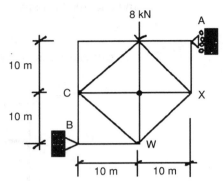

Exhibit 2.3

2.4 The force in member DE in Exhibit 2.4 is
  a. 11.2 kN tensile
  b. 11.3 kN compressive
  c. Zero
  d. None of the above

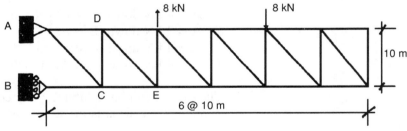

Exhibit 2.4

2.5 The forces in members FB and GH in Exhibit 2.5 are
  a. FB = 2.5 kN tensile, GH = 11.66 kN tensile
  b. FB = zero kN tensile, GH = 11.66 kN tensile
  c. FB = zero kN tensile, GH = 11.66 kN compressive
  d. FB = zero kN tensile, GH = 10.00 kN tensile

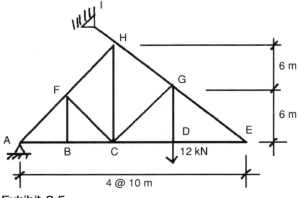

Exhibit 2.5

SOLUTIONS

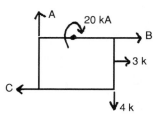

Exhibit 2.1a FBD whole structure

2.1 c.

$$\Sigma V = 0 \text{ (positive upward)}$$
$$+A - 4 = 0$$
$$A = 4 \text{ upward}$$

$$\Sigma M_c = 0 \text{ (positive clockwise)}$$
$$+20 + B(8) + 3(4) + 4(12) = 0$$
$$B = -10$$
$$B = 10 \text{ leftward}$$

$$\Sigma H = 0 \text{ (positive rightward)}$$
$$-C(-)10 + 3 = 0$$
$$C = -7$$
$$C = 7 \text{ rightward}$$

2.2 a.

$$\Sigma M_c = 0 \text{ (positive clockwise)}$$
$$+9 \times 12 - P(12) = 0$$
$$P = 9k$$

2.3 d.

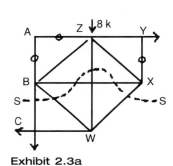

Exhibit 2.3a

Note: $XY = 0 = ZA = AB$.

$$\Sigma M_c = 0 \text{ (positive clockwise)}$$
$$+8(10) + R_Y(20) = 0$$
$$R_Y = 4 \leftarrow$$
So $YZ = 4$ comp.
ΣH gives $C_H = 4$ to the right
ΣV gives $C_V = 8$ upward

So

$$CB = 8c$$
$$CW = 4c$$

2.4 c.

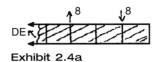

Exhibit 2.4a

FBD shown:

$$\Sigma V = 0$$
$$+8 - 8 + DE(0.7) = 0$$
$$DE = 0$$

2.5 b.

Joint B gives FB = 0

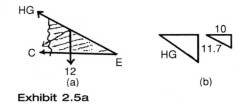

Exhibit 2.5a

FBD shown in Exhibit 2.5a:

$$\Sigma M_c = 0 \text{ (positive clockwise)}$$

$$+12 \times 10 - \left(\frac{10}{11.7} HG\right) \times 6$$

$$-\left(\frac{6}{11.7} HG\right) \times 10 = 0$$

$$120 - 5.13(HG) - 5.13(HG) = 0$$

$$HG = +\frac{120}{10.26} = 11.7$$

CHAPTER 3

Hydraulics and Hydro Systems

Bruce E. Larock

OUTLINE

MANNING EQUATION 23

HAZEN-WILLIAMS EQUATION 25

REFERENCES 25

This chapter will review the two equations that are likely to be found in the FE examination. A more thorough presentation may be found in *Civil Engineering: License Review* or in selected references at the end of this chapter.

MANNING EQUATION

The Manning equation for the average velocity V in a steady open channel flow is

$$V = \frac{K}{n} R^{2/3} S^{1/2}$$

With SI units $K = 1.0$; with English units $K = 1.49$. The hydraulic radius is $R = A/P$, with A = flow cross-sectional area and P = wetted perimeter (i.e., the length of the interface along the fluid/solid boundary containing it). The slope S of the energy line is equal to the amount of head lost in a channel section of length L, or $S = h_L/L$. A short table of values for n, the Manning roughness factor, follows:

Channel Surface	Roughness Value, n
Concrete, finished	0.012
Concrete, gunite	0.019
Clay, vitrified sewer	0.014
Rubble masonry	0.025
Concrete, mortar	0.013
Concrete, troweled	0.013
Gravel, clean	0.025

Although the Manning equation is intended primarily for use in open channels, it is sometimes also used for pressurized flow in pipes.

Example 3.1

A rectangular open channel lined with rubble masonry is 5 m wide and laid on a slope of 0.0004. If the depth of uniform flow is 3 m, compute the discharge in m³/s.

Solution

From the table, the Manning roughness is approximately $n = 0.025$. The area and wetted perimeter are

$$A = (5)(3) = 15 \text{ m}^2$$
$$P = 5 + 2(3) = 11 \text{ m}$$
$$R = A/P$$

The Manning equation now gives

$$Q = AV = A\left(\frac{1}{n}\right)R^{2/3}S^{1/2}$$

$$= (15)\left(\frac{1}{0.025}\right)(15/11)^{2/3}(0.0004)^{1/2}$$

$$= 14.76 \text{ m}^3/\text{s}$$

Example 3.2

A gunite concrete trapezoidal channel with 1:2 side slopes, shown in Exhibit 1, conveys 60 m³/s on a slope $S_o = 0.0005$. Compute the depth of uniform flow.

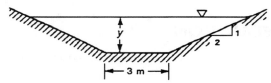

Exhibit 1

Solution

In SI units the Manning equation is

$$Q = (1/n)\, AR^{2/3}\, S_o^{2/3}$$

From the table the appropriate roughness coefficient is $n = 0.019$. Inserting the given information leads to

$$AR^{2/3} = 5.10 \tag{a}$$

in which

$$A = 3y + 2y^2$$
$$P = 3 + 2\sqrt{5}y \quad \text{and} \quad R = A/P$$

Equation (a) must now must be solved by successive trial. It is convenient to use a table in doing so:

Trial	y (m)	A (m²)	P (m)	R	$R^{2/3}$	$AR^{2/3}$ 5 5.10?
1	1.5	9.00	9.71	0.93	0.95	8.55
2	1.0	5.00	7.47	0.67	0.77	3.85
3	1.15	6.10	8.14	0.75	0.83	5.06
4	1.16	6.17	8.19	0.75	0.83	5.12

The normal depth, that is, the depth of uniform flow, is $y = 1.16$ m.

HAZEN-WILLIAMS EQUATION

Many empirical formulas for pipe friction have been developed over the past century. These formulas are usually based on tests involving the flow of water under fully turbulent conditions and are not normally reliable for use with other fluids. One relatively widely used such formula is the Hazen-Williams equation, which is

$$V = 0.849 \, CR^{0.63} \, S^{0.54}$$

for SI units. For English units, replace 0.849 with 1.318. The Hazen-Williams coefficient C ranges from approximately 140 for very smooth and straight pipes to 120 for smooth masonry to 100 or less for old cast iron pipe. The other factors in the equation are defined as they were for the Manning equation.

Example 3.3

If 0.01 m³/s of water flows through a new 100 mm clean cast iron pipe ($C = 130$), determine the head loss in 1000 m of this pipe.

Solution

The discharge $Q = VA$ with

$$A = \frac{\pi}{4}(0.1)^2 = 0.00785 \, \text{m}^2$$

Hence $V = 0.01/0.00785 = 1.273$ m/s
In this case the hydraulic radius is

$$R = \frac{\pi \frac{D^2}{4}}{\pi D} = \frac{D}{4} = \frac{0.1}{4} = 0.025 \, \text{m}$$

The Hazen-Williams equation then yields

$$V = 1.273 = 0.849 \, (130)(0.025)^{0.63} \, S^{0.54}$$
$$S = 0.0191 = h_L/L = h_L/1000 \quad \text{and} \quad h_L = 19.1 \, \text{m}$$

REFERENCES

Chow, V. T. *Open Channel Hydraulics*. McGraw-Hill, New York, 1959.
Henderson, F. M. *Open Channel Flow*. Macmillan, New York, 1966.
Street, R. L., Watters, G. Z., and Vennard, J. K. *Elementary Fluid Mechanics*, 7th Ed. Wiley, New York, 1996.
White, F. M. *Fluid Mechanics*, 3rd Ed. McGraw-Hill, New York, 1994.

CHAPTER 4

Structural Steel and Reinforced Concrete Design

Alan Williams

OUTLINE

ELASTIC DESIGN OF STEEL BEAMS 27
Bending Stresses ■ Shear Stress

COMPRESSION MEMBERS 31

TENSILE STRESS 34

STRENGTH DESIGN PRINCIPLES FOR REINFORCED CONCRETE MEMBERS 35
Flexure of Reinforced Concrete Beams ■ Deflection Requirements ■ Shear in Reinforced Concrete Members ■ Reinforced Concrete Columns

REFERENCES 44

PROBLEMS 45

SOLUTIONS 45

ELASTIC DESIGN OF STEEL BEAMS

The allowable stress on flexural members depends on the shape of the section and the bracing used to prevent lateral instability (Newman, 1995). Sections are classified as compact, noncompact, or slender in accordance with the criteria given in AISC Table B5.1. Most rolled W shapes qualify as compact sections. The exceptions are indicated in the AISC Tables of Properties by denoting the value of the yield stress F'_y and $F''_y\,'$ at which a particular shape becomes noncompact.

Bending Stresses

The allowable bending stress for compact symmetrical shapes is given by AISC Equation (F1-1) as

$$F_b = 0.66F_y$$

when the maximum unbraced length of the compression flange does not exceed the smaller of

$$L_c = \frac{76b_f}{\sqrt{F_y}} \quad \text{or} \quad \frac{20{,}000}{(dF_y/A_f)}$$

where b_f = flange width and A_f = compression flange area.

When the unbraced length of a compact shape exceeds L_c but is less than the greater of

$$L_u = r_T\sqrt{102{,}000C_b/F_y} \quad \text{or} \quad \frac{20{,}000C_b}{(dF_y/A_f)}$$

the allowable bending stress is given by AISC Section F1.3 as

$$F_b = 0.60\,F_y$$

where r_T = radius of gyration of the compression flange plus one-third of the web area. The bending coefficient C_b is defined in Section F1.3 and Table 6 as

$$C_b = 1.75 + 1.05\,M_1/M_2 + 0.3(M_1/M_2)^2 \quad \text{but not more than 2.3}$$

The value of C_b conservatively may be taken as unity. Examples of the derivation of C_b are illustrated in Figure 4.1. Values of L_c and L_u are tabulated in the AISC Beam Tables, assuming a value for C_b of unity.

In general, the allowable bending stress for noncompact symmetrical shapes is given by AISC Equation (F1-5) as

$$F_b = 0.60F_y$$

when the maximum unbraced length of the compression flange does not exceed L_c.

The AISC Selection Tables and Beam Tables tabulate the allowable beam resisting moments and uniformly distributed loads for W, M, S, C and MC shapes that are braced at the appropriate values of L_c or L_u. Adequate bracing is assumed to be provided by a restraint with a capacity of 1 percent of the force in the compression flange, in accordance with AISC Section G4. Note that metric steel tables are not available, so it is not possible to metricate the steel section.

When the unbraced length of the compression flange exceeds L_u, the allowable bending stress for both compact and noncompact shapes is given by the larger value from AISC Equations (F1-6), (F1-7), and (F1-8), but it may not exceed $0.60F_y$. When the unbraced length is less than

$$l = r_T\sqrt{510{,}000C_b/F_y} = L_1$$

the applicable equation is (F1-6), and the allowable stress is

$$F_b = F_y\left(0.667 - F_y l^2/1{,}530{,}000 r_T^2 C_b\right)$$

When the unbraced length equals or exceeds L_1, the applicable AISC equation is (F1-7), and the allowable stress is

$$F_b = \frac{170{,}000C_b r_T^2}{l^2}$$

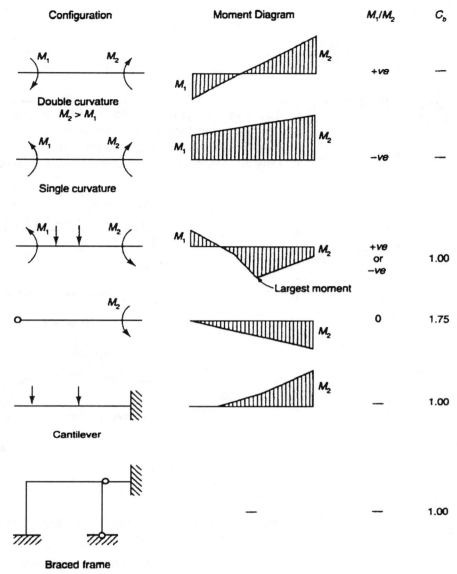

Figure 4.1 Derivation of C_b

Equation (F1-8) is independent of l/r_T and gives an allowable stress of

$$F_b = \frac{12,000 C_b A_f}{ld}$$

These expressions may be solved by calculator, or the allowable beam resisting moment may be obtained from AISC Moment Charts, which are based on a value of unity for C_b and are conservative for larger values of C_b.

Compact sections, solid rectangular sections bent about their weak axes, and solid round or square bars have an allowable stress given by AISC Equation (F2-1) of

$$F_b = 0.75 F_y$$

Noncompact sections bent about their weak axes have an allowable stress given by AISC Equation (F2-2) of

$$F_b = 0.60 F_y$$

Example 4.1

The W18 × 60 grade A36 beam shown in Exhibit 1 is laterally supported throughout its length. Determine whether the beam is adequate to support the applied loads indicated. The relevant properties of the beam are $S_x = 108$ in^3, $F_y = 36$ kips/in^2, and allowable bending stress $F_b = 0.66 F_y$.

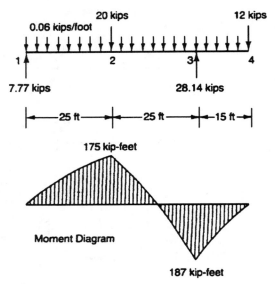

Exhibit 1

Solution

The bending moments acting on the beam from the applied loads and beam self-weight are shown in Exhibit 1.

$$M_x = \text{maximum moment} = 187 \text{ kip-ft}$$

$$f_b = \text{maximum bending stress} = M_x/S_x = 187 \times 12/108 = 20.78 \text{ kips/in}^2$$

The allowable stress is

$$F_b = 0.66 F_y = 0.66 \times 36 = 23.76 \text{ kips/in}^2$$

Hence, the W18 × 60 is adequate.

Shear Stress

The allowable shear stress, based on the overall beam depth, is given by AISC Equation (F4-1) as

$$F_v = 0.40 F_y$$

provided

$$h/t_w \leq 380/\sqrt{F_y}$$

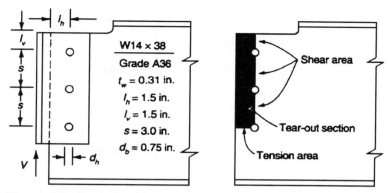

Figure 4.2 Block shear in a coped beam

and the actual shear stress is determined by

$$f_v = V/dt_w$$

where h = clear distance between the flanges, t_w = web thickness, d = overall depth of beam, and V = applied shear force.

When the end of the beam is coped, failure occurs by block shear, or web tear-out, which is a combination of shear along a vertical plane and tension along a horizontal plane. The resistance to block shear is given by

$$V_B = A_v F_v + A_t F_t$$

where A_v = net shear area, A_t = net tension area, F_v = allowable shear stress = $0.30F_u$ from AISC Equation (J4-1), and F_t = allowable tensile stress = $0.50F_u$ from AISC Equation (J4-2).

From Figure 4.2, where $d_h = d_b + 0.0625$ in.

$$A_v = t_w(l_v + 2s - 2.5d_h) = 0.31[1.5 + 6 - 2.5(0.75 + 0.0625)] = 1.70 \text{ in}^2$$
$$A_t = t_w(l_h - 0.5d_h) = 0.31(1.5 - 0.5 \times 0.8125) = 0.34 \text{ in}^2$$

For grade A36 steel, $F_u = 58$ kips per square inch.

The resistance to block shear is then

$$V_B = A_v F_v + A_t F_t = 1.7 \times 0.30 \times 58 + 0.34 \times 0.50 \times 58 = 39.44 \text{ kips}$$

COMPRESSION MEMBERS

The allowable stress in an axially loaded compression member is dependent on the slenderness ratio, which is defined in AISC Section E2 as Kl/r, where r = the governing radius of gyration, Kl = effective length of the member, K = effective-length factor, and l = unbraced length of the member.

The value of the effective-length factor depends on the restraint conditions at each end of the column. AISC Table C-C2.1 specifies effective-length factors for well-defined standard conditions of restraint, and these are illustrated in Figure 4.3. Values are indicated for ideal and practical end conditions, allowing for the fact that full fixity may not be realized. These values may be used only in simple cases when the tabulated end conditions are approached in practice.

End Restraints	Ideal K	Practical K
Fixed at both ends	0.5	0.65 
Fixed at one end, pinned at the other end	0.7	0.8 
Pinned at both ends	1.0	1.0 
Fixed at one end with the other end fixed in direction but not held in position	1.0	1.2 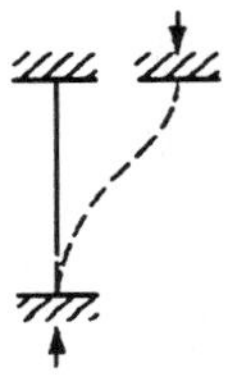
Pinned at one end with the other end fixed in direction but not held in position	2.0	2.0 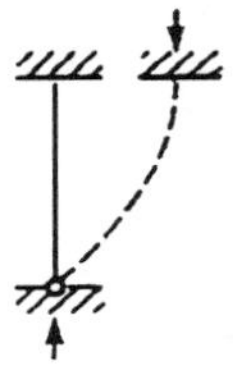
Fixed at one end with the other end free	2.0	2.1 

Figure 4.3 Effective-length factors

For compression members in a plane truss, AISC Section C-C2 specifies an effective-length factor of 1.0. For load-bearing web stiffeners on a girder, AISC Section K1.6 specifies an effective-length factor of 0.75. For columns in a rigid frame that is adequately braced, AISC Section C2.1 specifies a conservative value for the effective-length factor of 1.0.

The failure of a short, stocky column occurs at the squash load, when the strut yields in direct compression. The allowable stress in the strut is obtained from AISC Equation (E2-1) as

$$F_a = 0.6F_y$$

As the slenderness ratio of the column is increased, the failure load reduces. The Euler elastic critical load is assumed to govern when the column stress equals half the yield stress. The critical slenderness ratio, corresponding to this limit, is given by AISC Equation (C-E2-1) as

$$C_c = \sqrt{2\pi^2 \frac{E}{F_y}}$$

When the slenderness ratio exceeds this value, the allowable stress is

$$F_a = 12\pi^2 E/23(Kl/r)^2$$

When the slenderness ratio does not exceed this value, the allowable stress may be obtained from the expression

$$F_a = \left[1 - (Kl/r)^2/2C_c^2\right]F_y/\left[5/3 + 3(Kl/r)/8C_c - (Kl/r)^3/8C_c^3\right]$$

In accordance with AISC Section B7, the slenderness ratio should preferably not exceed 200.

Example 4.2

The W14 × 120 Grade A36 column shown in Exhibit 2 is fixed at the base and unbraced about the x-axis. About the y-axis the column is held in position, at the top and at mid-height, but it is not fixed in direction. Determine whether the column is adequate to support an axial load of 500 kips.

The relevant properties of the W 14 × 120 are $A = 35.3$ in^2, $r_x = 6.24$ in, $r_y = 3.74$ in, $E_s = 29{,}000$ ksi, $F_y = 36$ ksi, $K_y = 0.8$, and $K_x = 2.1$.

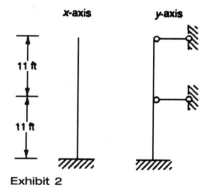

Exhibit 2

Solution

From Figure 4.3, the slenderness ratio about the y-axis is given by

$$Kl/r_y = 0.8 \times 11 \times 12/3.74 = 28.2$$

The slenderness ratio about the x-axis is given by

$$Kl/r_x = 2.1 \times 22 \times 12/6.24 = 88.8, \text{ which governs}$$

$$C_c = \sqrt{2\pi^2 \times 29{,}000/36} = 126.1 > kl/r$$

The allowable axial stress is

$$F_a = [1 - 88.8^2/(2 \times 126.1^2)]36/[5/3 + 3 \times 88.8/(8 \times 126.1) \\ - 88.8^3/(8 \times 126.1^3)] = 14.34 \text{ ksi}$$

The axial stress from the imposed loading is

$$f_a = P/A = 500/35.3 = 14.16 \text{ kips/in}^2 < F_a.$$ The column is adequate.

TENSILE STRESS

In determining the capacity of a connection in direct tension, as shown in Figure 4.4, allowance must be made for the effective areas of the members and their method of attachment. To prevent excessive elongation of the members, which may lead to instability of the whole structure, AISC Section D1 limits the maximum tensile force on the connection to

$$P_t = 0.6 F_y A_g$$

where A_g = gross area of the member = bt.

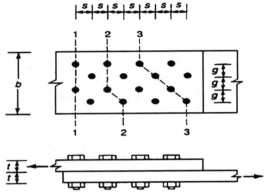

Figure 4.4 Net area of tension member

To prevent fracture of the member at the section of weakest effective net area, AISC Section D1 limits the maximum tensile force on the connection to

$$P_t = 0.5 F_u A_e$$

where A_e = effective net area. The effective net area is defined in AISC Section B2 as

$$A_e = t(b - 2d_h) \qquad \text{for the section 1-1 in Figure 4.4}$$

where d_h = specified diameter of hole = d_b + 1/8 in., where d_b = bolt diameter;

$$A_e = t(b - 3d_h + s^2/4g) \qquad \text{for section 2-2 in Figure 4.4}$$

where s = longitudinal pitch and g = transverse gage;

$$A_e = t(b - 4d_h + 3s^2/4g) \qquad \text{for section 3-3 in Figure 4.4}$$

To account for the effects of eccentricity and shear lag in rolled structural shapes connected through only part of their cross-sectional elements, the effective net area is given by AISC Equation (B3-1) as

$$A_e = U A_n$$

where A_n = net area of the member, U = 0.90 for I-sections with $b_f \geq 2d/3$ and with not less than three bolts in the direction of stress, U = 0.85 for all other shapes with not less than three bolts in the direction of stress, and $U = 0.75$ for all shapes with only two bolts in the direction of stress.

In addition, AISC Section B3 specifies that $A_e \leq 0.85 A_g$.

STRENGTH DESIGN PRINCIPLES FOR REINFORCED CONCRETE MEMBERS

The basic requirement of designing for strength is to ensure that the design strength of a member is not less than the required ultimate strength. The latter consists of the service-level loads multiplied by appropriate load factors, and this is defined in ACI Equations[2] (9-1), (9-2), and (9-3) as

$$U = 1.4D + 1.7L$$
$$U = 0.75(1.4D + 1.7L + 1.7W)$$
$$U = 0.9D + 1.3W$$

where D = dead load, L = live load, and W = wind load.

The design strength of a member consists of the theoretical ultimate strength of the member—the **nominal strength**—multiplied by the appropriate strength reduction factor, ϕ. Thus

$$\phi(\text{nominal strength}) \geq U$$

ACI Section 9.3 defines the reduction factor as $\phi = 0.90$ for flexure, $\phi = 0.85$ for shear and torsion, $\phi = 0.75$ for compression members with spiral reinforcement, $\phi = 0.70$ for compression members with lateral ties, and $\phi = 0.70$ for bearing on concrete.

Flexure of Reinforced Concrete Beams

The nominal strength of a rectangular beam, with tension reinforcement only, is derived from the assumed ultimate conditions shown in Figure 4.5. ACI Section 10.2.7.1 specifies an equivalent rectangular stress block in the concrete of $0.85 f'_c$, with a depth of

$$a = A_s f_y / 0.85 f'_c b = \beta_1 c$$

where c = depth to neutral axis and β_1 = compression zone factor, given in ACI Section 10.2.7.3.

From Figure 4.5, the nominal strength of the member is derived as

$$M_n = A_s f_y d (1 - 0.59 \rho f_y / f'_c)$$

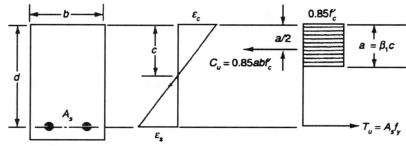

Figure 4.5 Member with tension reinforcement only

where $\rho = A_s/b_d$ = reinforcement ratio, $\phi M_n = 0.9 M_n$ = design strength, and $\phi M_n \geq M_u$ = applied factored moment. This expression may also be rearranged to give the reinforcement ratio required to provide a given factored moment, M_u, as

$$\rho = 0.85 f_c' \left[1 - \sqrt{1 - K/0.383 f_c'}\right]/f_y$$

where $K = M_u/bd^2$.

These expressions may be readily applied using standard calculator programs (Williams, 1996) and tables (American Concrete Assoc., 1985; Ghosh and Domel, 1992). For a balanced strain condition, the maximum strain in the concrete—and in the tension reinforcement—must simultaneously reach the values specified in ACI Section 10.3.2 as

$$\varepsilon_c = 0.003 = \text{concrete strain}$$
$$\varepsilon_s = f_y/E_s = \text{steel strain}$$

The balanced reinforcement ratio is given as $\rho_b = 0.85 \times 600 \beta_1 f_c'/f_y(600 + f_y)$.

In accordance with ACI Section 10.3.3, the maximum allowable reinforcement ratio to ensure a ductile flexural failure with adequate warning of impending failure is

$$\rho_{max} = 0.75 \rho_b$$

The maximum allowable reinforcement area is $A_{max} = bd\rho_{max}$.

The maximum allowable design strength of a singly reinforced member is

$$M_{max} = \phi A_{max} f_y d(1 - 0.59 \rho_{max} f_y/f_c')$$

In accordance with ACI Sections 10.5.1. and 10.5.2, the minimum allowable reinforcement ratio is given by

$$\rho_{min} = 1.4/f_y$$

with the exception that the minimum reinforcement provided need not exceed one-third more than required by analysis.

Example 4.3

A reinforced concrete beam, with an overall depth of 400 mm, an effective depth of 350 mm, and a width of 300 mm, is reinforced with Grade 400M bars and has a concrete cylinder strength of 21 MPa. Determine the area of tension reinforcement required for the beam to support a superimposed live load of 15 kN/m over an effective span of 6 m.

Solution

The weight of the beam is $w_D = 0.4 \times 0.3 \times 23.5 = 2.82$ kN/m.

The dead load moment is given by $M_D = w_D P^2/8 = 2.82 \times 6^2/8 = 12.7$ kNm.

The live load moment is given by $M_L = w_L P^2/8 = 15 \times 6^2/8 = 67.5$ kNm.

The factored moment is $M_u = 1.4 M_D + 1.7 M_L = 1.4 \times 12.7 + 1.7 \times 67.5 = 132.5$ kNm.

The moment factor is $K = M_u/bd^2 = 132.5 \times 10^6/(300 \times 350^2) = 3.605$ MPa.

The reinforcement ratio required to provide a given factored moment M_u is

$$\rho = 0.85 f'_c \, [1 - (1 - K/0.383 f'_c \,)^{0.5}]/f_y$$
$$= 0.85 \times 21 \{1 - [1 - 3.605/(0.383 \times 21)]^{0.5}\}/400$$
$$= 0.011$$

The minimum allowable ratio is

$$\rho_{min} = 1.4/f_y = 1.4/400 = 0.0035 < \rho \qquad \text{Satisfactory}$$

The maximum allowable reinforcement ratio is

$$\rho_b = 0.75 \times 0.85 \times 600 \beta_1 \, f'_c \,/f_y(600 + f_y)$$
$$= 0.75 \times 0.85 \times 600 \times 0.85 \times 21/400(600 + 400)$$
$$= 0.017 > \rho \qquad \text{Satisfactory}$$

Hence the section size is adequate.

The reinforcement area required is $A_s = \rho_b d = 0.011 \times 300 \times 350 = 1155$ mm^2

When the applied factored moment exceeds the maximum design strength of a singly reinforced member that has the maximum allowable reinforcement ratio, compression reinforcement and additional tensile reinforcement must be provided, as shown in Figure 4.6. The difference between the applied factored moment and the maximum design moment strength of a singly reinforced section is $M_r = M_u - M_{max}$ = residual moment.

The additional area of tensile reinforcement required is

$$A_T = M_r/\phi f_y (d - d') = A'_s f'_s / f_y$$

The depth of the stress block is

$$a = f_y A_{max}/0.85 f'_c b$$

The depth of the neutral axis is

$$c = a/\beta_1$$

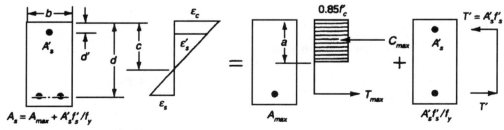

Figure 4.6 Member with compression reinforcement

The stress in the compression reinforcement is given by (Williams, 1996)

$$f'_s = 600(1 - d'/c) \le f_y$$

The required area of compression reinforcement is given by

$$A'_s = M_r/\phi f'_s (d - d')$$

The total required area of tension reinforcement is

$$A_s = A_{max} + A'_s f'_s / f_y$$

The maximum allowable reinforcement ratio is given by Section 10.3.3 as

$$\rho'_{max} = 0.75\rho_b + A'_s f'_s / bd f_y$$

In order to analyze a given member with compression reinforcement, an initial estimate of the neutral axis depth is required. The total compressive force in the concrete and compression reinforcement is then compared with the tensile force in the tension reinforcement. The initial estimate of the neutral axis depth is then adjusted until these two values are equal.

The conditions at ultimate load in a flanged member, when the depth of the equivalent rectangular stress block exceeds the flange thickness, are shown in Figure 4.7. The area of reinforcement required to balance the compressive force in the flange is given by

$$A_{sf} = C_f / f_y = 0.85 f'_c h_f (b - b_w) / f_y$$

The corresponding design moment strength is

$$M_f = \phi A_{sf} f_y (d - h_f/2)$$

The residual moment is $M_r = M_u - M_f$. The required reinforcement ratio to provide the residual moment is

$$\rho_w = 0.85 f'_c \left[1 - \sqrt{1 - 2K_w / 0.9 \times 0.85 f'_c} \right] / f_y$$

where $K_w = M_r / b_w d^2$.

The corresponding reinforcement area is

$$A_{sw} = b_w d \rho_w$$

The total reinforcement area required is

$$A_s = A_{sf} + A_{sw}$$

The total reinforcement ratio is

$$\rho = A_s / bd$$

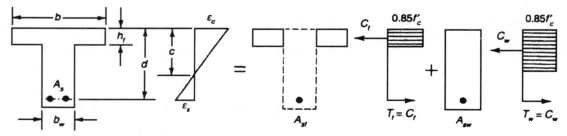

Figure 4.7 Flanged member with tension reinforcement

The maximum allowable reinforcement ratio is given by ACI Section 10.3.3 as

$$\rho'_{max} = 0.75 b_w (\rho_b + \rho_f)/b$$

where $\rho_b = 0.85 \times 600 \beta_1 f_c''/f_y (600 + f_y)$, and $\rho_f = A_{sf}/b_w d$.

Example 4.4

The reinforced concrete beam shown in Exhibit 3 is reinforced with Grade 400M bars at the positions indicated and has a concrete cylinder strength of 21 MPa. The beam carries a superimposed load of 30 kN/m run over an effective span of 6 m. Determine the areas of tension and compression steel required.

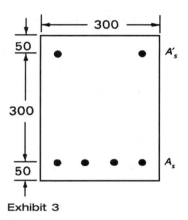

Exhibit 3

Solution

The weight of the beam is $w_D = 0.4 \times 0.3 \times 23.5 = 2.82$ kN/m. The dead load moment is given by $M_D = w_D l^2/8 = 2.82 \times 6^2/8 = 12.7$ kNm. The live load moment is given by $M_L = w_L l^2/8 = 30 \times 6^2/8 = 135.0$ kNm. The factored moment at midspan is obtained from ACI Equation (9-1) as

$$M_u = 1.4 M_D + 1.7 M_L = 1.4 \times 12.7 + 1.7 \times 135.0 = 247 \text{ kNm}$$

The compression zone factor is given by ACI Section 10.2.7 as $\beta_1 = 0.85$. The maximum allowable reinforcement ratio for a singly reinforced beam is

$$\begin{aligned}\rho_{max} &= 0.75 \times 0.85 \times 600 \times \beta_1 f_c''/f_y (600 + f_y) \\ &= 0.75 \times 0.85 \times 600 \times 0.85 \times 21/400(600 + 400) \\ &= 0.017\end{aligned}$$

The maximum reinforcement area for a singly reinforced beam is

$$A_{max} = bd \rho_{max} = 300 \times 400 \times 0.017 = 2040 \text{ mm}^2$$

The maximum design moment of a singly reinforced section is

$$\begin{aligned}M_{max} &= \phi A_{max} f_y d (1 - 0.59 \rho_{max} f_y / f_c') \\ &= 0.9 \times 2040 \times 400 \times 135(1 - 0.59 \times 0.017 \times 400/21)10^6 = 208 \text{ kNm}\end{aligned}$$

The residual moment is given by

$$M_r = M_u - M_{max} = 39 \text{ kNm}$$

The depth of the stress block is

$$\begin{aligned}a &= f_y A_{max}/0.85 f'_c b \\ &= 400 \times 2040/(0.85 \times 21 \times 300) \\ &= 152 \text{ mm}\end{aligned}$$

The neutral axis depth is

$$c = a/\beta_1 = 152/0.85 = 179 \text{ mm}$$

The stress in the compression reinforcement is

$$\begin{aligned}f'_s &= 600(1 - d'/c) = 600 - (1 - 50/179) \\ &= 400 \text{ MPa maximum}\end{aligned}$$

The required area of compression reinforcement is

$$\begin{aligned}A'_s &= M_r/\phi_s f'_s(d - d') = 10^6 \times 39/0.9 \times 400 \times 300 \\ &= 361 \text{ mm}^2\end{aligned}$$

The total required area of tension reinforcement is

$$\begin{aligned}A_s &= A_{max} + A'_s f'_s/f_y = 2040 + 361 \\ &= 2401 \text{ mm}^2\end{aligned}$$

Deflection Requirements

Allowable deflections are given in ACI Table 9.5(b). For reinforced concrete members not supporting deflection-sensitive construction, the allowable deflection may be deemed satisfied if the minimum thickness requirements of ACI Tables 9.5(a) and 9.5(c) are adopted.

For other conditions, the short-term deflection may be computed from the effective moment of inertia given by ACI Equation (9-7) and indicated in Figure 4.8 as

$$I_e = (M_{cr}/M_a)^3 I_g + [1 - (M_{cr}/M_a)^3] I_{cr}$$

where I_g = moment of inertia of gross concrete section, neglecting reinforcement = $bh^3/12$; I_{cr} = moment of inertia of the cracked transformed section = $b(kd)^3/3 + nA_s(d - kd)^2$; k = neutral axis depth ratio at service load for a singly reinforced section = $[2\rho n + (\rho n)^2]^{0.5} - \rho n$; $n = E_s/E_c$ = modular ratio; $\rho = A_s/bd$ = reinforcement ratio; M_a = maximum moment in the member; M_{cr} = cracking moment of the section $2f_r I_g/h$; and $f_r = 0.62\sqrt{f'_c}$ = modulus of rupture of normal-weight concrete.

When the member is subjected to long-term loads, creep and shrinkage produce additional deflection, which can be determined by multiplying the short-term deflection by the factor

$$\lambda = \xi/(1 + 50\rho')$$

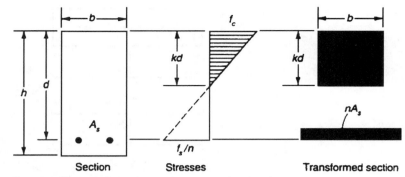

Figure 4.8 Reinforced member at service load

where ξ = time-dependent factor for sustained loads given in ACI Section 9.5.2.5, and $\rho' = A'_s/b_w d$ = reinforcement ratio for compression reinforcement.

Shear in Reinforced Concrete Members

The factored shear force acting on a member, in accordance with ACI Equations (11-1) and (11-2), is resisted by the combined design shear strength of the concrete and shear reinforcement. Thus, the factored applied shear is given by

$$V_u = \phi V_c + \phi V_s$$

where $\phi = 0.85$ = strength reduction factor for shear given by ACI Section 9.3.2.3; $V_c = 0.166\sqrt{f'_c}b_w d$ = nominal shear strength of normal weight concrete from ACI (11-3); $V_s = A_v f_y d/s$ = nominal shear strength of shear reinforcement from ACI Equation (11-17); A_v = area of shear reinforcement; and s = spacing of shear reinforcement, specified in ACI Section 11.5.4 as

$$s \leq d/2 \text{ or } 610 \text{ mm when } V_s \leq 0.33\sqrt{f'_c}b_w d,$$
$$s \leq d/4 \text{ or } 305 \text{ mm when } 0.33\sqrt{f'_c}b_w d < V_s \leq 0.66\sqrt{f'_c}b_w d$$

The dimensions of the section or the strength of the concrete must be increased, in accordance with ACI Section 11.5.6.8, to ensure that $V_s > 0.66\sqrt{f'_c}b_w d$.

As specified in ACI Section 11.5.5, a minimum area of shear reinforcement is required when

$$V_u > \phi V_c/2 = V_{u(\min)}$$

and the minimum area of shear reinforcement is given by ACI Equation (11-14) as

$$A_{v(\min)} = 0.34 b_w s/f_y$$

When the support reaction produces a compressive stress in the member, ACI Section 11.1.3.1 specifies that the critical factored shear force is the force that is acting at a distance equal to the effective depth from the face of the support. Figure 4.9 summarizes the shear provisions of the Code, and Figure 4.10 illustrates the design principles involved.

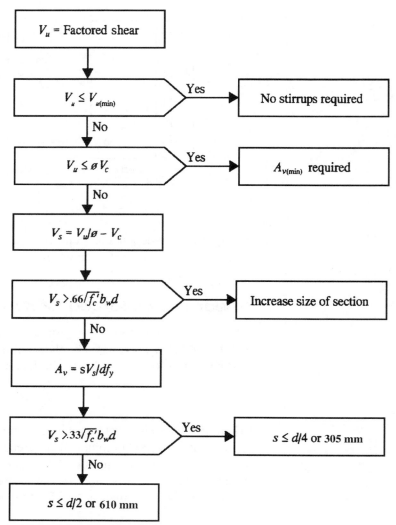

Figure 4.9 Flow chart for design for shear

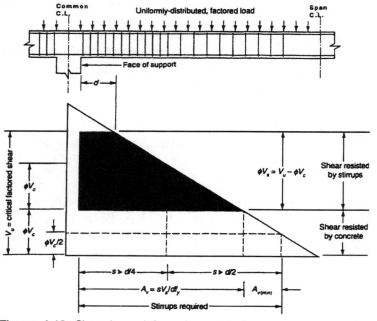

Figure 4.10 Shear in a reinforced concrete beam

Example 4.5

A reinforced concrete beam, which has an effective depth of 600 mm and a width of 500 mm, is reinforced with Grade 400M bars and has a compressive strength of 21 MPa. The factored shear force, at four locations on the beam, is (a) 800 kN, (b) 400 kN, (c) 100 kN, and (d) 500 kN. Determine the required spacing, at each location, for No. M10 stirrups with two or four vertical legs as suitable.

Solution

The design shear strength provided by the concrete is

$$\phi V_c = 0.166 \, \phi b_w d(f_c')0.5$$
$$= 0.166 \times 0.85 \times 500 \times 600(21)^{0.5}/1000$$
$$= 194 \text{ kN}$$

(a)
The design shear strength required from the shear reinforcement is

$$\phi V_s = V_u - \phi V_c = 800 - 194 = 606 \text{ kN}$$

Since $\phi V_s < 4 \times \phi V_c$ the concrete section is adequate.
Since $\phi V_s > 2 \times \phi V_c$ the maximum stirrup spacing is given by

$$s = d/4 = 600/4 = 150 \text{ mm}$$

The area of shear reinforcement required is

$$A_v/s = \phi V_s/\phi df_y = 606 \times 10^6/(0.85 \times 600 \times 400)$$
$$= 2970 \text{ mm}^2/\text{m}$$

The required spacing of stirrups with four No. M10 vertical legs is

$$s = 4 \times 100/2.97 = 135 \text{ mm} < 150 \text{ mm} \quad \text{Satisfactory}$$

(b)
The design shear strength required from the shear reinforcement is

$$\phi V_s = V_u - \phi V_c = 400 - 194 = 206 \text{ kN}$$

Since $\phi V_s < 2 \times \phi V_c$, the maximum stirrup spacing is given by

$$s = d/2 = 600/2 = 300 \text{ mm}$$

Reinforced Concrete Columns

Reinforced concrete columns may be classified as either short columns or long columns. For a short column, the axial load carrying capacity may be illustrated by reference to the column shown in Figure 4.11.

The theoretical design axial load strength at zero eccentricity is given by ACI Section R10.3.5 as

$$\phi P_0 = \phi[0.85 f_c'(A_g - A_{st}) + A_{st} f_y]$$

where

ϕ = strength reduction factor specified in Section 9.3
= 0.75 for compression members with spiral reinforcement and
= 0.70 for compression members with lateral ties

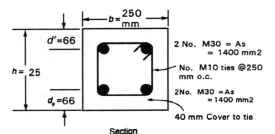

Figure 4.11 Compression in a short tied column

f'_c = concrete cylinder strength = 21 MPa
f_y = reinforcement yield strength = 400 MPa
A_g = gross area of the section = 62,500 mm^2
A_{st} = reinforcement area = 2800 mm^2

Then

$$\phi P_0 = 0.7[0.85 \times 21(62,500 - 2800) + 2800 \times 400]/1000 = 1530 \text{ kN}$$

To account for accidental eccentricity, ACI Section 10.3.5 requires a spirally reinforced column to be designed for a minimum eccentricity of approximately $0.05h$, which gives a maximum design axial load strength at zero eccentricity, in accordance with ACI Equation (10-1), of

$$\phi P_{n\max} = 0.85 \phi P_0$$

In the case of a column with lateral ties, a minimum eccentricity of approximately $0.10h$ is specified, which gives a maximum design axial load strength at zero eccentricity, in accordance with ACI Equation (10-2), of

$$\phi P_{n\max} = 0.80 \phi P_0 = 0.80 \times 1530 = 1224 \text{ kN}$$

REFERENCES

American Concrete Institute. *Building Code Requirements and Commentary for Reinforced Concrete* (ACI 318-89). Detroit, MI, 1995.

American Concrete Institute. *Design Handbook in Accordance with the Strength Design Method*. Detroit, MI, 1985.

Ghosh, S. K., and Domel, A. W. *Design of Concrete Buildings for Earthquake and Wind Forces*. Portland Cement Association, Skokie, IL, 1992.

Newnan, D. G., *Civil Engineering License Review*. Engineering Press, Austin, TX, 1995.

Williams, A. *Design of Reinforced Concrete Structures*. Engineering Press, Austin, TX, 1996.

PROBLEMS

In each of the following problems, the short column shown in Exhibit 4.1 supports a steel column and base plate with the indicated loads. The short column is 450 mm square and is reinforced with Grade 400M deformed bars. Concrete compressive strength is 14 MPa. The short column is subjected to axial load only and is effectively braced against sidesway by the floor slab.

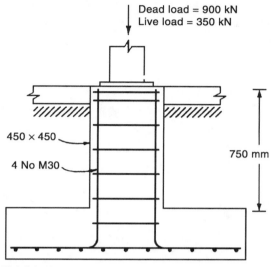

Exhibit 4.1

4.1 The design axial load strength of the short column is given most nearly by
 a. 1360 kN c. 1760 kN
 b. 1560 kN d. 1960 kN

4.2 The factored applied load is given most nearly by
 a. 1850 kN c. 1950 kN
 b. 1900 kN d. 2000 kN

4.3 The minimum allowable reinforcement area in the short column is most nearly
 a. 2000 mm² c. 2200 mm²
 b. 2100 mm² d. 2300 mm²

SOLUTIONS

4.1 d. For a column with lateral ties, the design axial load strength is given by Eq. (10.2) as

$$\phi P_n = 0.80\ \phi\ [0.85 f'_c (A_g - A_{st}) + f_y A_{st}]$$
$$= 0.8 \times 0.7[0.85 \times 14(202{,}500 - 2800) + 400 \times 2800]/1000$$
$$= 1958\text{ kN}$$

4.2 a. The applied ultimate load is

$$P_u = 1.4 \times 900 + 1.7 \times 350$$
$$= 1855\text{ kN}$$

4.3 a. The minimum allowable reinforcement area, in accordance with ACI Section 10.9.1, is

$$\begin{aligned}\rho_{\min} &= 0.01\, A_g \\ &= 0.01 \times 202{,}500 \\ &= 2025 \text{ mm}^2\end{aligned}$$

CHAPTER 5

Wastewater and Solid Waste Treatment

Kenneth J. Williamson

OUTLINE

WASTEWATER FLOWS 47

SEWER DESIGN 47
Hydraulics of Sewers

WASTEWATER CHARACTERISTICS 48
Oxygen Demand

WASTEWATER TREATMENT 49
Process Analysis ■ Physical Treatment Processes ■ Biological Treatment Processes ■ Disinfection

PROBLEMS 58

SOLUTIONS 59

WASTEWATER FLOWS

Wastewater flows comprise domestic and industrial wastewaters, infiltration, inflow, and storm water. Most modern sanitary sewers are separated from storm water systems, so these flows are treated separately. Modern sewers are constructed so that inflow rates are assumed to be negligible.

Domestic flows are determined from water use rates, with typical values being 0.57 m³ per day per capita. Industrial and infiltration flows vary widely. Specific data are required based upon industry and production rates.

SEWER DESIGN

Hydraulics of Sewers

Sewers are designed as open channels, usually with a circular cross section. Flows in sewers are modeled using Manning's equation (see Chapter 3) as

$$V = \frac{1}{n} R^{2/3} S^{1/2} \tag{5.1}$$

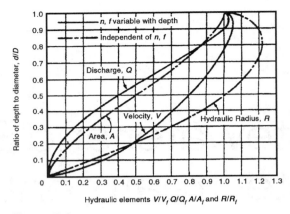

Figure 5.1

where
  V = velocity (m/s)
  n = Manning coefficient, 0.013
  R = hydraulic radius (m)
  S = slope of hydraulic grade line

The relation between the hydraulic radius and the flow depth for a circular cross section is a complex relationship; as a result, flows in partially full sewers are calculated using nomographs like that in Figure 5.1.

Design calculations usually involve knowing the slope of the sewer (S) and the partially full flow rate, Q. The nomograph is solved by assuming a pipe diameter (D), solving for the full flow rate, Q_f. From the ratio of Q/Q_f, the ratio of partial depth to pipe diameter (d/D) is obtained from the nomograph, from which the depth of flow is calculated. Sanitary sewers are designed to carry the peak flow with a depth of flow from one-half to full.

WASTEWATER CHARACTERISTICS

Wastewater can be characterized by a large number of parameters. Typical concentrations for design are given in Table 5.1. Per capita loading rates for 5-day biochemical oxygen demand and total suspended solids are 100 and 120 g/capita-d, respectively.

Table 5.1 Typical composition of domestic wastewater

Constituent	Concentration (mg/L)
Dissolved solids	700
Volatile dissolved solids	300
Total suspended solids	220
Volatile suspended solids	135
5-day biochemical oxygen demand	200
Organic nitrogen	15
Ammonia nitrogen	25
Total phosphorus	8

Oxygen Demand

Theoretical Oxygen Demand
Theoretical oxygen demand (ThOD) is a value calculated as the oxygen required to convert organic compounds in the wastewater to carbon dioxide and water. To determine the ThOD, the chemical formula of the waste must be known.

Chemical Oxygen Demand
Chemical oxygen demand (COD) is an empirical parameter representing the equivalent amount of oxygen that would be used to oxidize organic compounds under strong chemical oxidizing conditions. The COD and the ThOD are approximately equal for most wastes.

Biochemical Oxygen Demand
Biochemical oxygen demand (BOD) is an empirical parameter representing the amount of oxygen that would be used to oxidize organic compounds by aerobic bacteria. The test involves seeding of diluted wastewater and measurement of the oxygen depletion after five days in 300-mL glass bottles. The 5-day BOD (BOD_5) is related to the ultimate BOD as

$$BOD_5 = BOD_L(1 - 10^{-kt}) \qquad (5.2)$$

where
BOD_L = the maximum BOD exerted after a long time of incubation (mg/L);
k = the base-10 BOD decay coefficient (d^{-1})

The BOD test is accomplished at 20°C; the decay coefficient can be converted to other temperatures as

$$k_T = k_{20}\theta^{(T-20)} \qquad (5.3)$$

where
k_T, k_{20} = BOD decay coefficient at temperature T and 20°C, respectively
θ = temperature correction coefficient, 1.056 for 20–30°C and 1.135 for 4–20°C

Nitrogenous Oxygen Demand
Nitrogenous oxygen demand results from the oxidation of ammonia to nitrate by nitrifying bacteria. The overall equation is

$$NH_4^+ + 2O_2 = NO_3^- + 2H^+ + 2H_2O \qquad (5.4)$$

Nitrogenous oxygen demand is not included in either the BOD_5 or the COD value.

WASTEWATER TREATMENT

Wastewater treatment in the United States is managed under the Federal Water Pollution Control Act Amendments of 1972. Design of wastewater treatment facilities focuses on the two parameters BOD and total suspended solids (TSS). The general approach is to link together a series of unit processes to remove BOD and TSS sequentially to meet the discharge requirements. The unit processes are typically chosen to minimize treatment costs.

Process Analysis

Types of Reactions

Most reactions in wastewater treatment are considered to be homogeneous. In homogeneous reactions, the reaction occurs throughout the liquid, and mass transfer effects can be ignored. Chlorination is an example of a homogeneous reaction. In a heterogeneous reaction, the reactants must be transferred to a reactive site and the products must be transferred away from the reactive site. Biological reactions in bacterial films are an example of heterogeneous reactions.

Reaction Rates

Chemical reactions have many different forms, although the most common form is the conversion of a single product to a single reactant:

$$A \rightarrow B \tag{5.5}$$

If the rate of the reaction is zero-order, then the rate of conversion of A is

$$\frac{dA}{dt} = -k_0 \tag{5.6}$$

where k_0 = the zero-order reaction coefficient (mg/L-d).

If the rate of the reaction is first-order, then the rate of conversion of A is

$$\frac{dA}{dt} = -k_1[A] \tag{5.7}$$

where k_1 = first-order reaction coefficient (d^{-1}).

Reactor Types

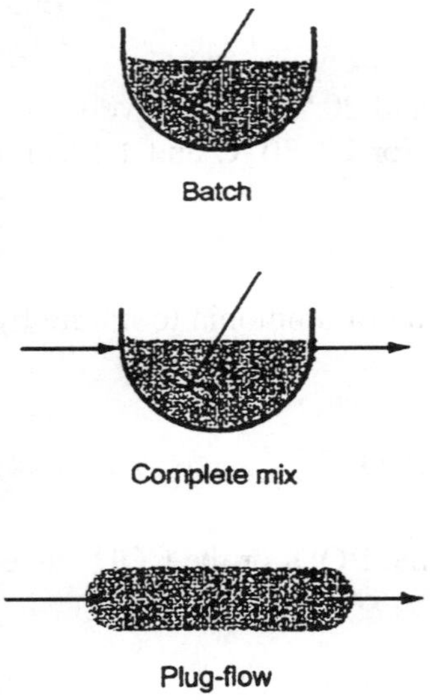

Three important reactor types are batch, complete-mix, and plug flow, as shown in Figure 5.2. Batch reactors have no influent or effluent and have a hydraulic detention time of t_b, which is the time between filling and emptying. Complete-mix reactors have constant influent and effluent flow rates and are mixed so that spatial gradients of concentration are near zero. They have a hydraulic detention time as

$$\theta_h = \frac{V}{Q} \tag{5.8}$$

where
- θ = hydraulic detention time (d)
- V = reactor volume (L)
- Q = influent and effluent flow rate (L/d)

Plug flow reactors have constant influent and effluent flow rates but lack internal mixing. The reactors tend to be long and narrow, and the transport of water occurs from one end to the other. Their hydraulic detention time is

$$t_r = \frac{L}{v_e} \tag{5.9}$$

where
- t_r = hydraulic detention time (d)
- L = length of reaction (m)
- v_e = liquid flow velocity (m/d)

Batch

Complete mix

Plug-flow

Figure 5.2

The application of mass balances to the three types of reactors, assuming either zero- or first-order reaction rates, results in the following equations for the effluent concentration of A at steady state:

Batch Reactor, Zero-Order

$$A = A_0 - k_0 t \tag{5.10}$$

Complete-Mix, Zero-Order

$$A = A_0 - k_0 \theta_h \tag{5.11}$$

Plug Flow, Zero-Order

$$A = A_0 - k_0 t_r \tag{5.12}$$

Batch Reactor, First-Order

$$A = A_0 e^{(-k_1 t_b)} \tag{5.13}$$

Complete-Mix, First-Order

$$A = \frac{A_0}{1 + k_1 \theta_k} \tag{5.14}$$

Plug Flow, First-Order

$$A = A_0 e^{-k_1 t_r} \tag{5.15}$$

Plug flow reactors will consistently give lower effluent concentrations as compared to complete-mix reactors for all reaction rates greater than zero.

Physical Treatment Processes

Sedimentation
Sedimentation is divided into four types: discrete (Type 1); flocculent (Type 2); zone (Type 3); and compression (Type 4).

Discrete Sedimentation
Discrete sedimentation is described in Chapter 7. In discrete sedimentation, the removal is determined by the terminal settling velocity. Settling velocities for small Reynolds numbers can be estimated by Stokes' law as

$$V_p = \frac{g(\rho_s - \rho_w)d^2}{18\mu} \tag{5.16}$$

where
V_p = particle settling velocity (m/s)
g = gravitational constant (9.8 m/s²)
ρ_p, ρ_w = density of particle and water, respectively (kg/m³)
μ = water viscosity (N-s/m²)

For particles with V_p less than the overflow rate of the sedimentation basin, V_c, the removal R is

$$R = \frac{V_p}{V_c} \qquad (5.17)$$

R is 1.0 for all particles with $V_p > V_s$.

Flocculent Sedimentation

Flocculent sedimentation is similar to Type 1 settling except that the particle size increases as the particle settles in the sedimentation reactor. Type 2 or flocculent settling occurs in coagulation/flocculation treatment for water and in the top of secondary sedimentation basins.

Removal under flocculent sedimentation is determined empirically. Data must be collected to give isolines for percent removal as shown in Figure 5.3. Given such data, the total removal is

$$R = \sum \left(\frac{\Delta h_n}{h_s}\right)\left(\frac{R_n + R_{n+1}}{2}\right) \qquad (5.18)$$

where the symbols are as illustrated in Figure 5.3.

Zone Settling

Zone settling occurs in the bottom of a secondary sedimentation basin, where the concentration of the solids increases to over 5000 mg/L. A schematic of a sedimentation basin coupled to a biological treatment reactor with cell recycle is shown in Figure 5.4. Under such conditions, the modeling of the zone sedimentation process is based upon the solids flux through the bottom section of the basin. The solids flux results from two components: gravity sedimentation and liquid velocity from the underflow.

$$SF = SF_g + SF_r \qquad (5.19)$$

where

SF, SF_g, SF_r = solids flux, solids flux due to gravity, and solids flux due to recycle, respectively (kg/m^2-d)

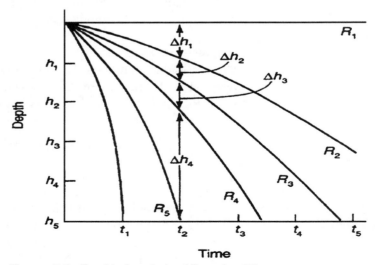

Figure 5.3 Empirical analysis of Type 3 settling

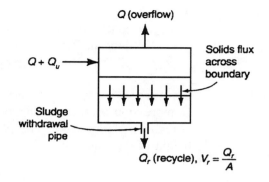

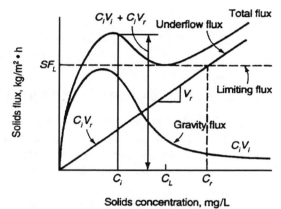

Figure 5.4 Schematic of secondary clarifier and empirical analysis of Type 3 settling

The resulting relationship is

$$\text{SF} = C_i V_i + C_i V_r \tag{5.20}$$

where
C_i = concentration of solids at a specified height in the basin (mg/L)
V_i = gravity settling velocity of sludge with concentration C_i (m/d)
V_r = downward fluid velocity from recycle (m/d)
    = Q_r/A_H
A_H = horizontal area of sedimentation basin

Equation (5.19) is graphed in Figure 5.4. Under a constant recycle flow and a defined sludge, the transport of solids through the bottom of the clarifier becomes limited by the limiting solids flux rate (SF_L) shown in Figure 5.4. Estimates of SF_L are obtained by graphical analysis as illustrated in Figure 5.4. The horizontal area of the sedimentation basin is then calculated as

$$A = \frac{(Q+Q_r)(C_{\text{infl}})}{\text{SF}_L} \tag{5.21}$$

or

$$A = \frac{Q_r C_r}{\text{SF}_L} \tag{5.22}$$

where
- C_{inf1} = concentration of TSS in influent to sedimentation basin
- C_r = concentration of TSS in recycle from sedimentation basin
- Q, Q_r = plant influent and recycle flow rates

Compression Sedimentation

Compression sedimentation involves the slow movement of water through the pores in a sludge cake. The process is modeled as an exponential decrease in height of the sludge cake as

$$H_t - H_\infty = (H_0 - H_\infty)e^{-i(t-t_0)} \tag{5.23}$$

where
- H_t, H_∞, H_0 = sludge height after time t, a long period, and initially
- i = compression coefficient.

Biological Treatment Processes

Biological treatment involves the conversion of organic and inorganic compounds by bacteria with subsequent growth of the organisms. The common biological treatment processes in wastewater treatment involve the removal of various substrates including BOD, ammonia, nitrate, and organic sludges.

Description of Homogeneous Processes

The rate of removal of a substrate is given as

$$r_{su} = -\frac{k \times S}{(K_s + S)} \tag{5.24}$$

where
- r_{su} = rate of substrate removal (mg/L-d)
- k = maximum substrate removal rate (mg/mg-d)
- S, K_s = substrate and half-velocity coefficient, respectively (mg/L)

The growth of organisms is given as

$$r_X = Y r_{su} - k_d X \tag{5.25}$$

where
- r_X = rate of growth of bacteria (mg/L-d)
- Y = substrate yield rate (mg/mg)
- k_d = bacterial decay coefficient (d^{-1})
- X = bacterial concentration, usually expressed as TVSS (mg/L)

Complete-Mix Reactor Without Cell Recycle

For the complete-mix reactor without cell recycle shown in Figure 5.5(a), the governing equations are

$$X = \frac{Y(S_0 - S)}{(1 + k_d \theta_h)} \tag{5.26}$$

$$S = \frac{K_s(1 + \theta_h k_d)}{\theta_h (YK - k_d) - 1} \tag{5.27}$$

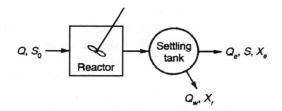

(a) Complete-Mix Without Cell Recycle

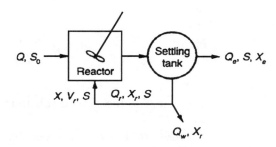

(b) Complete-Mix With Cell Recycle

Figure 5.5 Schematic of complete-mix reactors

Complete-Mix Reactor with Cell Recycle
For the complete-mix reactor with cell recycle shown in Figure 5.5(b), the governing equations are

$$\frac{1}{\theta_c} = Y\left(\frac{kSX}{S+K_s}\right) - k_d \tag{5.28}$$

where θ_c = solids retention time (d);

$$\frac{F}{M} = \frac{S_0}{\theta_h X} \tag{5.29}$$

where F/M = food-to-microorganism ratio (mg/mg-d); and

$$X = \frac{\theta_c Y(S_0 - S)}{\theta_h(1 + k_d \theta_c)} \tag{5.30}$$

$$S = \frac{K_s(1 + \theta_c k_d)}{\theta_c(Yk - k_d) - 1} \tag{5.31}$$

Process design proceeds from selection of a solids retention time (θ_c) as the controlling parameter. The volume of the reactor is determined from Equation (5.30) with $\theta_h = V/Q$. The waste sludge flow rate is computed from

$$\theta_c = \frac{VX}{Q_w X_w + Q_e X_e} \tag{5.32}$$

where

Q_w, Q_e = sludge waste and effluent flow rates (L/d)
X_w = sludge waste concentration (X if from reaction, X_r if from the clarifier) (mg/L)
X_e = effluent TSS concentration (mg/L), usually 20 mg/L

Plug Flow Reactor with Cell Recycle

The cell concentration in a plug flow reactor can be estimated using Equation (5.30). The effluent substrate concentration can be estimated from

$$\frac{1}{\theta_c} = \frac{Yk(S_0 - S)}{(S_0 - S) + \left(1 + \frac{Q_r}{Q}\right) K_s \ln(S_i/S)} - k_d \qquad (5.33)$$

where S_i = the diluted substrate concentration entering the reactor (mg/L).

Description of Heterogeneous Processes

Trickling filters are designed using an empirical equation as

$$A = Q \left(\frac{-\ln(S_{\text{eff}}/S_{\text{infl}})}{k_1 D}\right)^2 \qquad (5.34)$$

where
A = horizontal area
k_1 = BOD decay coefficient
D = filter depth

Disinfection

Disinfection in wastewater treatment is almost universally accomplished using chlorine gas. The objective of disinfection is to kill bacteria, viruses, and amoebic cysts. Effluent standards for secondary treatment require fecal coliform levels of less than 200 and 400 per 100 mL for 30-day and 7-day averages, respectively.

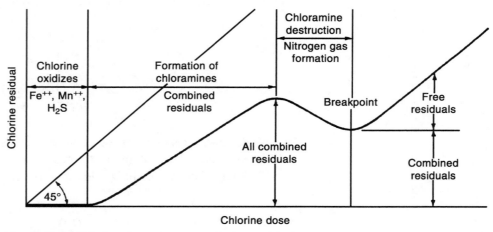

Figure 5.6 Chlorine dosage versus chlorine residual

Modeling of disinfection is described in Chapter 7. The major difference between chlorination of water supplies and that of wastewater is the reaction of chlorine with ammonia to produce chloramines as

$$NH_3 + HOCl \rightarrow NH_2Cl + H_2O$$

Chloramines can undergo further oxidation to nitrogen gas with their subsequent removal.

The stepwise oxidation of various compounds is shown in Figure 5.6. As chlorine is initially added, it reacts with easily oxidized compounds such as reduced iron, sulfur, and manganese. Further addition results in the formation of chloramines, collectively termed **combined chlorine residual**. Further addition of chlorine results in destruction of the chloramines and ultimately to the formation of free chlorine residuals as $HOCl$ and OCl^-. The development of free chlorine residual is termed **breakpoint chlorination**.

PROBLEMS

5.1 What is the required slope for a 0.3-m diameter circular sewer as shown in Exhibit 5.1, with a flow of 0.014 m^3/s and a minimum velocity of 0.6 m/s?

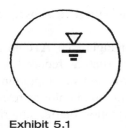

Exhibit 5.1

5.2 Compute the carbonaceous and nitrogenous oxygen demands for a waste with a chemical formula of $C_5H_7NO_2$.

5.3 A lake with a volume of 5×10^6 m^3 has a freshwater flow of 20 m^3/s. A waste is dumped into the lake at a rate of 50 g/s with a decay rate of 0.2/d. What is the steady-state concentration? Assume that the lake is completely mixed.

5.4 A waste with a flow of 2.8 L/s is discharged to a small stream with a flow of 141 L/s. The waste has a 5-d BOD of 200 mg/L ($k = 0.2$/d). What is the BOD after one day's travel in the stream?

5.5 A sedimentation basin has an overflow of 3 ft/hr. The influent wastewater has a particle distribution as follows:

Percent of Particles	Settling Velocity (m/hr)
20	0.30 to 0.61
30	0.61 to 0.91
50	0.91 to 1.22

Determine the total removal in the sedimentation basin.

5.6 Disinfection with chlorine is known to be a first-order reaction. The first-order decay rate under a given concentration of chlorine is measured as 0.35/hr. The flow rate is 12,000 L per hour, and the desired removal of organisms is from 10×10^6/100mL to less than 1/100mL. Determine
a. Volume of reactor required assuming the use of a complete-mix reactor
b. Volume of reactor required assuming the use of plug flow reactor

5.7 A Type 3 settling test was conducted on a waste activated sludge. The results were as follows:

MLSS (mg/L)	Settling Velocity (m/hr)
4000	2.4
6000	1.2
8000	0.55
10,000	0.31
20,000	0.06

Determine the limiting solids flux if the concentration of the recycled solids is 10,000 mg/L.

5.8 Find the effluent soluble BOD and reactor cell concentration for an aerobic complete-mix reactor with no recycle.

$$k = 10 \text{ mg/mg-d}$$
$$K_s = 50 \text{ mg/L}$$
$$k_d = 0.10/\text{d}$$
$$Y = 0.6 \text{ mg/mg}$$
$$S_0 = 200 \text{ mg BOD/L}$$
$$\theta_h = \theta_c = 2\text{d}$$

SOLUTIONS

5.1
$$D_f = 0.30 \text{ m}$$
$$Q = 0.014 \text{ m}^3/\text{s}$$
$$V = 0.6 \text{ m/s}$$

Need V/V_f or Q/Q_p or A/A_f or R/R_f

$$A_f = \frac{\pi D_f^2}{4} = 0.071 \text{ m}^2$$

$$A = \frac{Q}{V} = 0.023 \text{ m}^2$$

$$\frac{A}{A_f} = 0.32$$

From nomograph (Exhibit 5.1a)

$$d/D_f = 0.35 \quad d = 0.105 \text{ m}$$
$$R/R_f = 0.75$$

$$R_f = \frac{\pi D_f^2/4}{\pi D} = \frac{D_f}{4} = 0.075 \text{ m}$$

$$R = (0.75)(0.075) = 0.056 \text{ m}$$

From Manning's equation:

$$V = \frac{1}{n} R^{2/3} S^{1/2}$$

$$S = \left(\frac{nV}{1.R^{2/3}}\right)^2$$

$$= \left(\frac{(0.013)(0.6 \text{ m/s})}{(1.)(0.056)^{2/3}}\right)^2$$

$$= 0.0028 \text{ m/m}$$

5.2 For ThOD:
$$C_5H_7NO_2 + 5O_2 \rightarrow 5CO_2 + NH_3 + 2H_2O$$

For NOD:
$$NH_3 + 2O_2 \rightarrow HNO_3 + H_2O$$

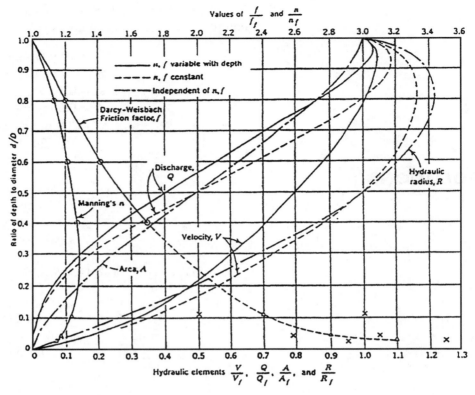

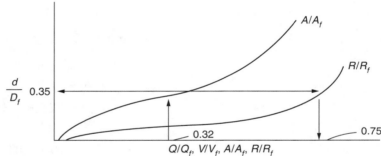

Exhibit 5.1a

For carbonaceous oxygen demand:

$$\frac{5 \text{ moles O}_2}{1 \text{ mole C}_5\text{H}_7\text{NO}_2} \times \frac{32 \text{ g O}_2}{\text{mole O}_2} \times \frac{1 \text{ mole C}_5\text{H}_7\text{NO}_2}{113 \text{ g}} = \frac{1.4 \text{ g O}_2}{\text{g C}_5\text{H}_7\text{NO}_2}$$

For nitrogenous oxygen demand:

$$\frac{2 \text{ moles O}_2}{1 \text{ mole C}_5\text{H}_7\text{NO}_2} \times \frac{32 \text{ g O}_2}{\text{mole O}_2} \times \frac{1 \text{ mole C}_5\text{H}_7\text{NO}_2}{113 \text{ g}} = \frac{0.57 \text{ g O}_2}{\text{g C}_5\text{H}_7\text{NO}_2}$$

5.3

$$C = \frac{S}{Q + kV}$$

$$= \frac{50 \text{ g/s}}{20 \text{ m}^3/\text{s} + (0.2/\text{d})(5 \times 10^6 \text{ m}^3)\left(\frac{1 \text{ d}}{8.64 \times 10^4 \text{ s}}\right)}$$

$$= 1.58 \text{ g/m}^3$$

5.4

$$BOD_L = \frac{BOD_5}{(1+e^{-5k})}$$

$$= \frac{20 \text{ mg/L}}{\left(1-e^{-\left(5d \times \frac{2}{d}\right)}\right)}$$

$$= 294 \text{ mg/L}$$

After mixing,

$$BOD_i = \frac{(294 \text{ mg/L})(2.8 \text{ L/s})}{141.6 \text{ L/s} + 2.8 \text{ L/s}}$$
$$= 5.76 \text{ mg/L}$$

Assume the stream is plug flow.

$$BOD = BOD_i e^{-kt}$$
$$= (5.76 \text{ mg/L})e^{-(0.21d)(1d)}$$
$$= 4.7 \text{ mg/L}$$

5.5

Particles	Avg V_p (cm/hr)	V_c (cm/hr)	R (%)
2.54–6.45	3.81	7.62	50
6.45–7.62	6.35	7.62	83
7.62–10.16	8.89	7.62	100

$$R_{total} = (0.2)(0.5) + (0.3)(0.83) + (0.5)(1.0)$$
$$= 0.85 \text{ or } 85\%$$

5.6

a.

$$\frac{A}{A_0} = \frac{1}{1+k_1 \frac{v}{Q}} = \frac{1}{10 \times 10^6}$$

$$1 + k_1 \frac{v}{Q} = 1 \times 10^7$$

$$V = \frac{(1 \times 10^7)(12,000 \text{ L/hr})}{0.35/\text{hr}}$$
$$= 3.4 \times 10^{11} \text{ L}$$

b.

$$\frac{A}{A_0} = e^{-kv/Q} = \frac{1}{10 \times 10^6}$$

$$-k\frac{v}{Q} = -16.1$$

$$V = \frac{-16.1(12,000 \text{ L/hr})}{0.35/\text{hr}}$$
$$= 5.5 \times 10^5 \text{ L}$$

5.7

MLSS (mg/L)	v_i (ft/hr)	$x_i v_i$ (mg-ft/L-hr)	v_i (cm/hr)	$x_i v_i$ (kg/m²-d)
4000	7.8	3.12×10^4	19.81	4.63
6000	3.8	2.28×10^4	9.65	3.11
8000	1.8	1.44×10^4	4.57	1.97
10,000	1.0	1.00×10^4	2.54	1.37
20,000	0.2	0.80×10^4	0.51	1.09

From graphical solution (see Exhibit 5.7)

$$\text{Limiting solids flux} = 7 \text{ kg/m}^2\text{-d}$$

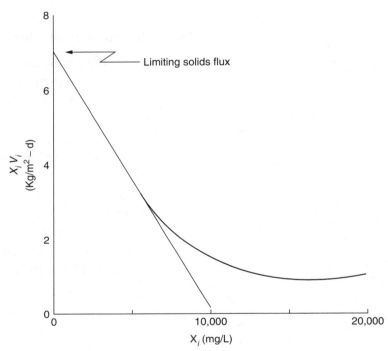

Exhibit 5.7

5.8

$$s = \frac{K_s(1+\theta_h k_d)}{\theta_h(Y_k - k_d) - 1}$$

$$= \frac{(50 \text{ mg/L})(1+(2 \text{ d})(0.10/\text{d}))}{(2 \text{ d})\left(\frac{0.6 \text{ mg}}{\text{mg}} \times \frac{10 \text{ mg}}{\text{mg-d}} - \frac{0.10}{\text{d}}\right)^{-1}}$$

$$= 5.7 \text{ mg/L}$$

$$X = \frac{4(S_0 - S)}{1 + k_d \theta_h} = \frac{\left(\frac{0.6 \text{ mg}}{\text{mg}}\right)(200 \text{ mg/L} - 5.7 \text{ mg/L})}{1 + (0.10/\text{d})(2 \text{ d})}$$

$$= 97 \text{ mg/L}$$

CHAPTER 6

Transportation Engineering

Robert W. Stokes

OUTLINE

HIGHWAY CURVES 63
Simple (Circular) Horizontal Curves ■ Vertical Curves

SIGHT DISTANCE 70
Sight Distance on Simple Horizontal Curves ■ Sight Distance on Vertical Curves

TRAFFIC CHARACTERISTICS 73

EARTHWORK 74

REFERENCES 76

PROBLEMS 78

SOLUTIONS 79

HIGHWAY CURVES

Simple (Circular) Horizontal Curves

The location of highway centerlines is initially laid out as a series of straight lines (tangent sections). These tangents are then joined by circular curves to allow for smooth vehicle operations at the design speed selected for the highway. Figure 6.1 shows the basic geometry of a simple circular curve. If the two tangents intersecting at PI are laid out, and the angle Δ between them is measured, only one other element of the curve need be known to calculate the remaining elements. The radius of the curve (R) is the other element most commonly used.

Circular Curve Formulas
The following symbols are defined as shown in Figure 6.1:

PC = Point of curvature (beginning of curve)
PI = Point of intersection
PT = Point of tangency (end of curve)
Δ = Intersection or central angle, degrees

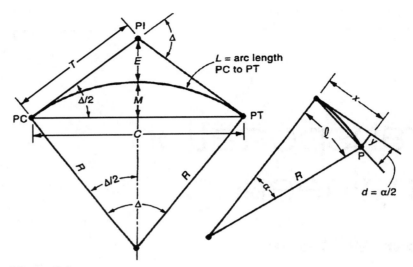

Figure 6.1

R = Radius of curve, m
L = Length of curve, m
E = External distance, m
M = Middle ordinate, m
C = Long chord, m
P = Length of arc between any two points on curve, m
α = Central angle subtended by arc P, degrees
d = Deflection angle for any arc length P, degrees
x = Distance along tangent from PC or PT to set any point P on curve, m
y = Offset (normal) from tangent at distance x to set any point P on curve, m

Then

$L = \Delta R/57.2958$
$T = R \tan \Delta/2$
$E = R(\sec \Delta/2 - 1) = R \operatorname{exsec} \Delta/2$
$M = R(1 - \cos \Delta/2) = R \operatorname{vers} \Delta/2$
$C = 2R \sin \Delta/2$
$R = (C^2 + 4M^2)/8M$ (formula for estimating R from field measurements)
$P = (\alpha \times R)/57.2958$
$d = \alpha/2 = 1718.873 \ P/R$ (in minutes)
   $= 28.64789 \ P/R$ (in degrees)

For any length x,

$$y = R - (R^2 - x^2)^{1/2}$$

For any length P,

$$X = R \sin \alpha$$
$$Y = R(1 - \cos \alpha) = R \operatorname{vers} \alpha$$

Example 6.1

Given the following horizontal curve data, determine the curve radius, R; length of curve, L; stationing (sta) of the PC; stationing of the PT; the long chord, C; and deflection angle, d, of a point on the curve 30 m ahead of the PC.

$$PI = \text{sta } 2 + 170.00 \text{ (km)}$$
$$\Delta = 41° 10'$$
$$T = 115.00 \text{ m}$$

Solution

$$PC \text{ sta} = PI \text{ sta} - T = 2170.00 - 115.00 = 2 + 055.00$$
$$R = T/\tan(\Delta/2) = 115.00/0.3755 = 306.22 \text{ m}$$
$$L = (\Delta R)/57.2958 = 220.02 \text{ m}$$
$$PT \text{ sta} = PC \text{ sta} + L = 2055.00 + 220.02 = 2275.02 \text{ m}$$
$$C = 2R \sin(\Delta/2) = 2(306.22)(0.3516) = 215.32 \text{ m}$$
$$d = (28.64789P)/R = [(28.64689)(30)]/306.22 = 2.81°$$

Horizontal Curve Design

The minimum radius of horizontal curvature is determined by the dynamics of vehicle operation and sight distance requirements. The minimum radius necessary for the vehicle to remain in equilibrium with respect to the incline of a horizontal curve is given by the following formula:

$$R = V^2/[127 \, (e + f)]$$

where V = vehicle speed in km/h, e = rate of superelevation ("banking") in m/m of width, and f = allowable side friction factor. The equation can be used to solve for the minimum radius for a given speed, or the maximum safe speed for a given radius.

Example 6.2

An existing horizontal curve has a radius of 235 m. What is the maximum safe speed on the curve? Assume $e = 0.08$ and $f = 0.14$.

Solution

$$R = V^2/[127 \, (e + f)]$$
$$V = [R[127 \, (e + f)]]^{0.5}$$
$$V = [235 \, [127 \, (0.08 + 0.14]]^{0.5} = 81 \text{ km/h}$$

Vertical Curves

The vertical-axis parabola (see Figure 6.2) is the geometric curve most commonly used in the design of vertical highway curves. The general equation of the parabola is

$$Y = aX^2 + bX + c$$

where the constant a is an indication of the rate of change of slope, b is the slope of the back tangent (g_1), and c is the elevation of the PVC.

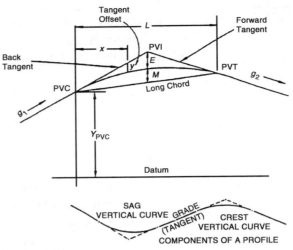

Figure 6.2

Vertical Curve Formulas

The following notation is defined as shown in Figure 6.2:

L = Length of curve (horizontal)
PVC = Point of vertical curvature
PVI = Point of vertical intersection
PVT = Point of vertical tangency
g_1 = Grade of back tangent (decimal)
g_2 = Grade of forward tangent (decimal)
a = Parabola constant
y = Tangent offset
E = Tangent offset at PVI
M = Middle ordinate
r = Rate of change of grade
K = Rate of curvature
A = Algebraic difference in grades
x = Horizontal distance from PVC to point on curve
x_m = Horizontal distance to min/max elevation on curve

Then

$A = g_2 - g_1$
$a = (g_2 - g_1)/2L = A/2L$
$r = (g_2 - g_1)/L = A/L$
$K = L/A$
$E = a\,(L/2)^2 = (AL)/8 = M$
$y = ax^2$
$x_m = -g_1/(2a) = -g_1 K$

Tangent elevation = $Y_{PVC} + g_1 x$
Curve elevation = $Y_{PVC} + g_1 x \pm ax^2$ (add for sag curves, subtract for crest curves)

Example 6.3

Determine the following for the sag vertical curve shown in Exhibit 1

a. Stationing and elevation of the PVC

b. Stationing and elevation of the PVT

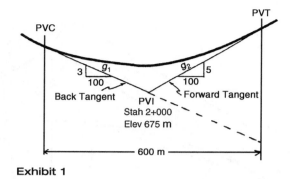

Exhibit 1

c. Elevation of the curve 100 m from the PVC
d. Elevation of the curve 400 m from the PVC
e. Stationing and elevation of the low point of the curve

Solution

a. The curve is symmetric about the PVI, therefore:

   Sta PVC = Sta PVI − L/2 = 2000 − 600/2 = 1 + 700.0
   Elev of the PVC = Elev PVI + g_1(L/2) = 675 + 0.03(300) = 684.0 m

b. Sta PVT = Sta PVI + L/2 = 2000 + 600/3 = 2 + 300.0

   Elev of the PVT = Elev PVI + g_2(L/2) = 675 + 0.05(300) = 690.0 m

c. The elevation of the curve at any distance x from the PVC is equal to the tangent offset + the tangent elevation. To calculate the tangent offset, it is first necessary to evaluate the constant a.

   $$a = (g_2 - g_1)/2L = [0.05 - (-0.03)]/[2(600)] = 0.000067$$

   Tangent offset at x = 100 m from the PVC = ax^2 = 0.000067 $(100)^2$ = 0.67 m
   Tangent elevation 100 m from the PVC = PVC elev + $g_1 x$
   = 684.0 m + (−0.03)100 = 681.0 m

   Curve elevation = tangent offset + tangent elev = 0.67 + 681.0 = 681.67 m

d. The elevation of the curve 400 m from the PVC can be calculated relative to the forward tangent or relative to the extension of the back tangent. Relative to the extension of the back tangent:

   Tangent offset = ax^2 = 0.000067 $(400)^2$ = 10.67 m
   Elev of the extended back tangent = PVC elev + $g_1 x$
   = 684.0 + (−0.03)400 = 672.0 m
   Curve elev = tangent offset + tangent elev = 10.67 + 672.0 = 682.67 m

e. The low point of the curve occurs at a distance x_m from the PVC.

   $x_m = -g_1/(2a) = -(-0.03)/[2(0.000067)] = 225$ m (from the PVC)

   Sta of the low point = Sta PVC + 225 = 1700 + 225 = 1 + 925.0
   The tangent offset at 225 m from the PVC = ax^2 = 0.000067$(225)^2$ = 3.38 m
   The tangent elevation at this distance = PVC elev + $g_1 x$
   = 684.0 + (−0.03)225 = 677.25 m
   The elevation of the low point = tangent offset + tangent elevation
   = 677.25 + 3.38 = 680.63 m

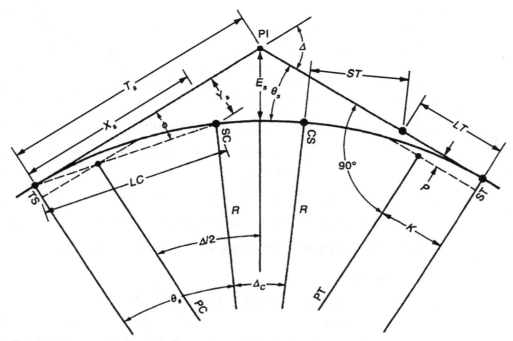

Figure 6.3

Transition (Spiral) Curves

The spiral curve is used to allow for a transitional path from tangent to circular curve, from circular curve to tangent, or from one curve to another that have substantially different radii. The minimum length of spiral curve needed to achieve this transition can be computed from the following formula.

$$P_s = V^3/(46.7RC)$$

where P_s = minimum length of spiral (m), V = design speed (km/hr), R = circular curve radius (m), and C = rate of increase of centripetal acceleration (m/s^3). The value $C = 0.6$ is generally used for highway curves.

Transition (Spiral) Curve Formulas

The basic geometry of the spiral curve is shown in Figure 6.3. The spiral is defined by its parameter A and the radius of the simple curve it joins. The product of the radius (r) at any point on the spiral and the corresponding spiral length (P) from the beginning of the spiral to that point is equal to the product of the radius (R) of the simple curve it joins and the total length (P_s) of the spiral as shown in the following equation:

$$rR = RR_s = \text{constant} = A^2$$

The following notation is defined as shown in Figure 6.3:

ℓ_s = total length of spiral from TS to SC
ℓ = spiral length from TS to any point on spiral
L_c = total length of circular curve
R = radius of circular curve
T_s = total tangent distance from the PI to the TS or ST
Δ = deflection angle between the tangents

θ_s = central angle of the entire spiral (spiral angle)
θ = central angle of any point on the spiral
Δ_c = central angle of simple curve
E_s = external distance
LC = long chord
LT = long tangent
ST = short tangent
X_s = tangent distance from TS to SC
x_s = tangent distance from TS to any point on the spiral
Y_s = tangent offset at the SC
y_s = tangent offset at any point on the spiral
k = simple curve coordinate (abscissa)
p = simple curve coordinate (ordinate)
ϕ = deflection angle at TS from the initial tangent to any point on the spiral
TS = tangent to spiral
SC = spiral to circular curve
CS = circular curve to spiral
ST = spiral to tangent
C_s = spiral deflection angle correction factor

Then

$$\theta_s = 28.648\,(P_s/R)$$
$$\theta = (P/P_s)^2 \theta_s$$
$$\phi = (\theta/3) - C_s$$
$$C_s \text{ (seconds)} = 0.0031\theta^3 + 0.0023\theta^5 \times 10^{-5}$$
$$Y_s = (\ell_s^2/6R) - (\ell_s^4/133R^3)$$
$$y_s = (\ell^2/6R) - (\ell^4/336R^3)$$
$$X_s = \ell_s - (\ell_s^3/40R^2)$$
$$p = Y_s - R(1 - \cos\theta_s)$$
$$k = X_s - R\sin\theta_s$$
$$T_s = (R + p)\tan(\Delta/2) + k$$
$$ST = Y_s/\sin\theta_s$$
$$LT = X_s - (Y_s/\tan\theta_s)$$

All distances are in meters and all angles are in degrees unless noted otherwise.

Example 6.4

A transition spiral is being designed to provide a gradual transition into a circular curve with a radius of 435 m and a design speed of 90 km/h. The PI of the curve is at Station 0 + 625.00, and the deflection angle between the tangents is 33.67 degrees. Determine the minimum length of spiral required and the stationing of the TS and the ST.

Solution

Assuming a value of $C = 0.6$, the minimum length of the spiral can be determined as follows.

$$\ell_s = V^3/(46.7RC) = (90)^3/(46.7 \times 435 \times 0.6) = 60 \text{ m}$$

To determine the stationing of the TS and the ST, proceed as follows:
Compute the spiral angle.

$$\theta_s = 28.648 \, (P_s/R) = 28.648 \, (60/435) = 4 \text{ degrees}$$

Compute Y_s and X_s.

$$Y_s \, (\ell_s^2/6R) - (\ell_s^4/336R^3) - = [(60)^2/(6)(435)] - [(60)^4/(336)(435)^3] = 1.379 \text{ m}$$

$$X_s = \ell_s - (\ell_s^3/40R^2) = 60 - [(60)^3/(40)(435)^2] = 59.972 \text{ m}$$

Compute p and k.

$$p = Y_s - R(1 - \cos\theta_s) = 1.379 - 435(1 - \cos 4°) = 0.319 \text{ m}$$
$$k = X_s - R \sin\theta_s = 59.972 - (435) \sin 4° = 29.628 \text{ m}$$

Compute the spiral tangent length.

$$T_s = (R + p) \tan(D/2) + k$$
$$= (435 + 0.319) \tan(33.67/2) + 29.628 = 161.349 \text{ m}$$

Compute the central angle of the circular curve.

$$\Delta_c = \Delta - 2\theta_s = 33.67 - 2(4) = 25.67 \text{ degrees}$$

Compute the length of the circular curve.

$$L = (\Delta_c R)/57.2958 = [25.67(435)]57.2958 = 194.891 \text{ m}$$

Compute the stationing of the TS.

$$\text{TS Sta} = \text{PI Sta} - T_s = 0 + 625 - 0 + 161.349 = 0 + 463.651$$

Compute the stationing of the ST.

$$\text{ST Sta} = \text{TS Sta} + L + 2y\ell_s$$
$$= 0 + 463.651 + 0 + 194.891 + 2(60) = 0 + 778.542$$

SIGHT DISTANCE

The ability of drivers to see the road ahead is of utmost importance in the design of highways. This ability to see is referred to as **sight distance** and is defined as the length of highway ahead that is visible to the driver.

Stopping sight distance is the sum of the distance traveled during the perception-reaction time and the distance traveled while braking to a stop. The stopping sight distance can be determined from the following equation.

$$S = 0.278Vt + V^2/[254 \, (f \pm g)]$$

where S = stopping sight distance (m), V = vehicle speed (km/hr), t = perception-reaction time (assumed to be 2.5 sec), f = coefficient of friction, and g = grade (*plus* for uphill, *minus* for downhill), expressed as a decimal.

Sight Distance on Simple Horizontal Curves

The required middle ordinates (M) for clear sight areas to satisfy stopping sight distance (S) requirements as a function of the radii (R) of simple horizontal curves can be determined using the following equation. The stopping sight distance is measured along the centerline of the inside lane of the curve.

$$M = R[1 - \cos(28.65 \, S/R)]$$

Example 6.5

A large outcropping of rock is located at M (middle ordinate) = 10 m from the centerline of the inside lane of a proposed highway curve. The radius of the proposed curve is 120 m. What speed limit would you recommend for this curve? Explain your answer. Assume $f = 0.28$, $e = 0.10$, perception-reaction time = 2.5 sec, and level grade.

Solution

To solve this problem you must determine whether sight distance or the radius of the curve controls the speed.
Determine the available sight distance.

$$M = R[1 - \cos(28.65\, S/R)]$$
$$S = (R/28.65) \cos^{-1}[(R-M)/R]$$
$$= (120/28.65)\cos^{-1}[(120-10)/120] = 98.67 \text{ m}$$

Determine the maximum safe speed on the curve as a function of stopping sight distance.

$$S = 0.278Vt + V^2/[254(f \pm g)]$$
$$98.67 = 0.278V(2.5) + V^2/[254(0.28)]$$
$$V^2 + 49.43V - 7017.41 = 0$$

and from the quadratic equation

$$V = 62.6 \text{ km/hr}$$

Determine the maximum safe speed on the curve as a function of the radius.

$$R = V^2/[127(e+f)]$$
$$V = [127R(e+f)]^{0.5} = [127(120)(0.10 + 0.28)]^{0.5} = 76.1 \text{ km/hr}$$

Because the sight distance is sufficient only for a speed of about 62.6 km/hr, set speed limit at 60 km/hr.

Sight Distance on Vertical Curves

Formulas for stopping and passing sight distances for *vertical curves* are summarized following. The formulas are based on an assumed driver eye height (h_1) of 1070 mm. The height of object (h_2) above the roadway surface is assumed to be 150 mm for stopping sight distance and 1300 mm for passing sight distance.

Formulas for Sight Distance on Vertical Curves

The following terms are defined as shown in Figure 6.4:

L = Length of vertical curve, m

A = Algebraic difference grades, %

S = Sight distance, m

K = Vertical curvature, L/A

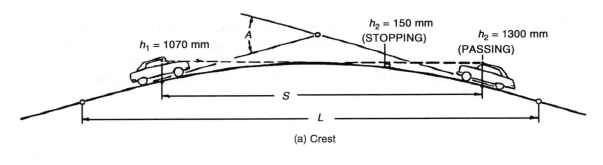

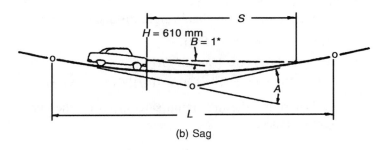

Figure 6.4

H = Headlight height, m

B = Upward divergence of light beam, degrees

Crest:

$$L = AS^2/100 \left(\sqrt{2h_i} + \sqrt{2h_2}\right)^2 \quad \text{when } S < L$$

The formula is different for when $S > L$, but does not apply to geometric design criteria.

$$\text{Stopping sight distance: } L = AS^2/404 \quad \text{or} \quad K = S^2/404$$
$$\text{Passing sight distance: } L = AS^2/946 \quad \text{or} \quad K = S^2/946$$

Sag:

$$L = AS^2/200\,(H + S \tan B) \quad \text{when } S < L$$

For passenger cars:

$$L = AS^2/(122 + 3.5\,S) \quad \text{or} \quad K = S^2/(122 + 3.5\,S)$$

Example 6.6

A crest vertical curve joins a +2% grade with a −2% grade. If the design speed of the highway is 95 km/hr, determine the minimum length of curve required. Assume $f = 0.29$ and the perception-reaction time = 2.5 sec. Also assume $S < L$.

Solution

Determine the stopping sight distance required for the design conditions.

$$S = 0.278Vt + V^2/[254\,(f \pm g)] = 0.278\,(95)(2.5) + (95)^2/[254(0.29 - 0.02)]$$
$$= 197.62 \text{ m}$$

(Note that the worst-case value for g is used.)

Determine the minimum length of vertical curve to satisfy the required sight distance.

$$L = AS^2/404 = [4(197.62)^2]/404 = 386.67 \text{ m}$$

TRAFFIC CHARACTERISTICS

The traffic on a roadway may be described by three general parameters: volume or rate of flow, speed, and density.

Traffic **volume** is the number of vehicles that pass a point on a highway during a specified time interval. Volumes can be expressed in daily or hourly volumes or in terms of subhourly rates of flow. There are four commonly used measures of daily volume.

1. **Average annual daily traffic** (AADT) is the average 24-hour traffic volume at a specific location over a full year (365 days).

2. **Average annual weekday traffic** (AAWT) is the average 24-hour traffic volume occurring on weekdays at a specific location over a full year.

3. **Average daily traffic** (ADT) is basically an estimate of AADT based on a time period less than a full year.

4. **Average weekday traffic** (AWT) is an estimate of AAWT based on a time period less than a full year.

Highways are generally designed on the basis of the **directional design hourly volume (DDHV)**.

$$\text{DDHV} = \text{AADT} \times K \times D$$

where AADT = average annual daily traffic (veh/day), K = proportion of daily traffic occurring in the design hour, D = proportion of design hour traffic traveling in the peak direction of travel.

For design purposes, K often represents the proportion of AADT occurring during the **thirtieth highest hour** of the year on rural highways, and the fiftieth highest hour of the year on urban highways.

Traffic volumes can also exhibit considerable variation within a given hour. The relationship between hourly volume and the maximum 15-minute rate of flow within the hour is defined as the **peak-hour factor (PHF)**.

$$\text{PHF} = V_h/(4V_{15})$$

where V_h = hourly volume (vph), V_{15} = maximum 15-minute rate of flow within the hour (veh).

The second major traffic stream parameter is **speed**. Two different measures of average speed are commonly used. **Time mean speed** is the average speed of all vehicles passing a point on the highway over a given time period. **Space mean speed** is the average speed obtained by measuring the instantaneous speeds of all vehicles on a section of roadway. Both of these measures can be calculated from a series of measured travel times over a measured distance from the following equations.

$$\mu_t = (\Sigma(d/t_i))/n$$
$$\mu_s = (nd)/\Sigma t_i$$

where μ_t = time mean speed (m/sec or km/hr), μ_s = space mean speed (m/sec or km/hr), d = distance traversed (m or km), n = number of travel times observed, t_i = travel time of ith vehicle (sec or hr).

Because the space mean speed weights slower vehicles more heavily than time mean speed does, it results in a lower average speed. The relationship between these two mean speeds is

$$\mu_t = \mu_s + \sigma/\mu_s$$

where σ = the variance of the space speed distribution.

Example 6.7

The following travel times were observed for four vehicles traversing a 1-km segment of highway.

Vehicle	Time (minutes)
1	1.6
2	1.2
3	1.5
4	1.7

Calculate the space and time mean speeds of these vehicles.

Solution

The space mean speed is

$$\mu_s = (nd)/\Sigma t_i = 4(1)/(1.6 + 1.2 + 1.5 + 1.7) = 0.67 \text{ km/min} = 40.0 \text{ km/hr}$$

The time mean speed is

$$\mu_t = \Sigma(d/t_i)/n = [(1/1.6) + (1/1.2) + (1/1.5) + (1/1.7)]/4$$
$$= 0.68 \text{ km/min} = 40.8 \text{ km/hr}$$

The third measure of traffic stream conditions, **density** (sometimes referred to as **concentration**), is the number of vehicles traveling over a unit length of highway at a given instant in time.

The general equation relating flow, density, and speed is

$$q = k\mu_s$$

where q = rate of flow (vph), μ_s = space mean speed (km/hr), and k = density (veh/km).

The relationship between density and flow shown in Figure 6.5 is commonly referred to as the "fundamental diagram of traffic flow." Note in the figure that (1) when density is zero, the flow is also zero; (2) as density increases, the flow also increases; and (3) when density reaches its maximum, referred to as jam density (k_j), the flow is zero.

EARTHWORK

One of the major objectives in evaluating alternative route locations is to minimize the amount of cut and fill. A common method of determining the volume of earthwork is the **average end area** method. This method is based on the assumption that the

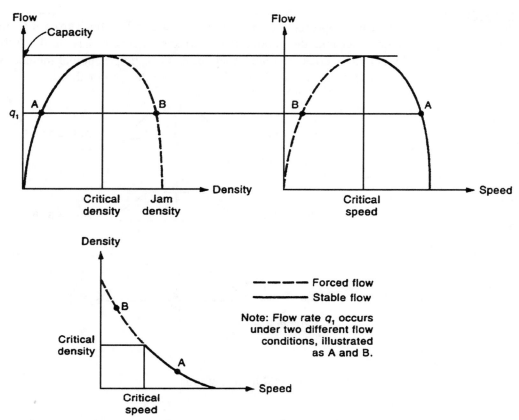

Figure 6.5 Relationships among speed, density, and rate of flow

volume between two consecutive cross sections is the average of their areas multiplied by the distance between them.

$$V = L(A_1 + A_2)/2$$

where V = volume (meters^3), A_1 and A_2 = end areas (m^2)

L = distance between cross sections (m)

In situations where there is a significant difference between A_1 and A_2, it may be advisable to calculate the volume as a pyramid:

$$V = (1/3)(\text{area of base})(\text{length})$$

where V is in cubic meters, the area of the base is in square meters, and the length is in meters.

The average end area and the pyramid methods provide "reasonably accurate" estimates of volumes of earthwork. When a more precise estimate of volume is desired, the **prismoidal formula** is frequently used:

$$V = L(A_1 + 4A_m + A_2)/6$$

where V = volume (m^3), A_1 and A_2 = end areas (m^2), and A_m = middle area determined by averaging corresponding *linear* dimensions (*not* the end areas) of the end sections (m^2).

When materials from cut sections are moved to fill sections, shrinkage factors (generally in the range of 1.10 to 1.25) are applied to the fill volumes to determine the quantities of fill required.

Example 6.8

Given the trapezoidal cross sections shown in Exhibit 2 from a temporary access ramp, determine the volume in cubic meters between stations 4 + 320 and 4 + 400 by (1) the average end area method, and (2) the prismoidal method. Comment on any differences in the results obtained from the two methods.

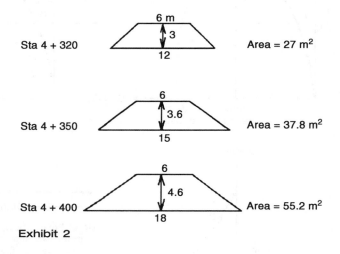

Exhibit 2

Solution

1. For the average end area method:

$$V = L/2 \, (A_1 + A_2)$$

The total volume is the sum of the volumes between stations 4 + 320 and 4 + 350 and between 4 + 350 and 4 + 400.

$$V = (30/2)(27.0 + 37.8) + (50/2)(37.8 + 55.2) = 3{,}297.9 \text{ m}^3$$

2. For the prismoidal method:

$$V = (L/6)(A_1 + 4A_m + A_2)$$

The middle areas (A_m) are calculated from the averages of the base, top, and height dimensions of the trapezoids.

The middle area between stations 4 + 320 and 4 + 350 is $[(6.0 + 13.5)/2] \times 3.3 = 32.2 \text{ m}^2$. The middle area between stations 4 + 350 and 4 + 400 is $[(6.0 + 16.5)/2] \times 4.1 = 46.1 \text{ m}^2$.

The total volume is the sum of the volumes between stations 4 + 320 and 4 + 350 and between 4 + 350 and 4 + 400.

$$V = (30/6)[27.0 + 4(32.2) + 37.8] + (50/6)[37.8 + 4(46.1) + 55.2] = 3{,}279.7 \text{ m}^3$$

Comment: The prismoidal volumes are theoretically more precise than the average end area volumes. The average end area method tends to overestimate volumes, as in this case.

REFERENCES

American Association of State Highway and Transportation Officials (AASHTO). *A Policy on Geometric Design of Highways and Streets.* Washington, DC, 1994.

Garber, N. J. and Hoel, L. A. *Traffic and Highway Engineering.* West Publishing Co., St. Paul, MN, 1988.

Hickerson, T. F. *Route Location and Design,* 5th ed. McGraw-Hill, New York, 1967.

Institute of Transportation Engineers (ITE). *Traffic Engineering Handbook,* 4th ed. Washington, DC, 1992.

Newnan, D. G., ed. *Civil Engineering License Review*, 12th ed. Engineering Press, Austin, TX, 1995.

Transportation Research Board (TRB). *Highway Capacity Manual,* Special Report 209, 3rd ed. TRB, Washington, DC, 1994.

PROBLEMS

6.1 For a simple horizontal curve with Δ = 12 degrees, R = 400 m, and PI at Station 0 + 241.782, the stationing of the PC and PT, respectively, would be most nearly
a. 0 + 197.7, 0 + 283.8
b. 0 + 197.7, 0 + 283.5
c. 0 + 158.0, 0 + 283.8
d. 0 + 195.6, 0 + 279.4

6.2 The minimum radius (m) for a simple horizontal curve with a design speed of 110 km/hr is most nearly (assume e = 0.08 and f = 0.14)
a. 425
b. 395
c. 550
d. 435

6.3 A vehicle hits a bridge abutment at a speed estimated by investigators as 25 km/hr. Skid marks of 30 m on the pavement (coefficient of friction, f = 0.35) followed by skid marks of 60 m on the gravel shoulder (f = 0.50) approaching the abutment are observed at the accident site. The grade is level. The initial speed (km/hr) of the vehicle was at least
a. 105
b. 90
c. 110
d. 95

For Questions 6.4 and 6.5, assume a + 3.9% grade intersects a + 1.1% grade at station 0 + 625.0 and elevation 305.0 m.

6.4 The minimum length (m) of the vertical curve for a design speed of 80 km/hr is most nearly (assume perception-reaction time = 2.5 sec, f = 0.30, and $S < L$)
a. 85
b. 130
c. 55
d. 110

6.5 The elevation (m) of the middle point of the curve is most nearly
a. 305.00
b. 305.45
c. 304.45
d. 304.55

6.6 The areas (in m²) of the cut sections of a proposed highway are shown below.

Stations	Area, m²
5 + 000	27.9
5 + 050	326.1
5 + 100	965.3
5 + 200	1651.8
5 + 300	2126.6

The total volume of cut (to the nearest cubic meter) between stations 5 + 000 and 5 + 100 is most nearly (use the average end area method):
a. 35,615
b. 34,865
c. 44,420
d. 41,135

6.7 A 11-m-wide vertical wall roadway tunnel is to be constructed using the cut and cover method. The cross sections to be excavated are shown below.

Station	Area, m²
1 + 000	20
1 + 100	30
1 + 130	35
1 + 200	40
1 + 260	45
1 + 300	40
1 + 400	35

The volume of earth in cubic meters to be excavated between stations 1 + 130 and 1 + 260 is most nearly (use the prismoidal method):
a. 5165
b. 5585
c. 7860
d. 2340

SOLUTIONS

6.1 b. Calculate the tangent length:

$$T = R \tan (\Delta/2) = 400 \tan (12/2) = 42.042 \text{ m}$$

Calculate the length of the curve:

$$L = (R\Delta)/57.2958 = 83.776 \text{ m}$$
$$\text{PC sta} = \text{PI sta} - T = 0 + 241.782 - 0 + 042.042 = 0 + 199.740$$
$$\text{PT sta} = \text{PC sta} + L = 0 + 199.740 + 0 + 083.776 = 0 + 283.516$$

6.2 d. $R = V^2/[127(e + f)] = (110)^2/[127(0.08 + 0.14)] = 433 \text{ m}$

6.3 a. The only known speed is the final collision speed of 25 km/hr. Therefore, consider the braking distance on the gravel first.

$$\text{Braking distance} = (V_0^2 - V^2)/[254\,(f \pm g)]$$

where V_0 and V represent the initial and final speeds, respectively.

$$\text{Braking distance (gravel)} = 60\text{ m} = (V_0^2 - 25^2)/(254)(0.5)$$

Solving for V_0, $V_0 = 90.8$ km/hr = speed at the beginning of the gravel and at the end of the pavement skid. Therefore, for the pavement skid,

$$\text{Braking distance (pavement)} = 30\text{ m} = (V_0^2 - 90.8^2)(254)(0.35)$$
$$\text{Solving for } V_0,\; V_0 = 104.5\text{ km/hr}$$

Comment: Note that this solution does not account for any speed reduction prior to the beginning of the skid marks on the pavement. Therefore, we conclude that the vehicle speed was *at least* 104.5 km/hr.

6.4 b. The required stopping sight distance is

$$S = 0.278\,Vt + V^2/[254(f \pm g)]$$
$$= 0.278(80)(2.5) + (80)^2/[254(0.30 + 0.011)] = 136.62\text{ m}$$

The minimum length of crest vertical curve to satisfy the required stopping sight distance is

$$L = (|A|S^2)/404 = [|(1.1 - 3.9)|(136.62)^2]/404 = 129.36\text{ m}$$

6.5 d. The midpoint offset is

$$E = (AL)/8 = [(0.011 - 0.039)(129.36)]/8 = -0.45\text{ m}$$

Therefore,

Elevation of the midpoint of the curve
= elevation of the PVI (305 m) − 0.45 m = 304.55 m

Note: The elevation of the midpoint could also be found using the basic properties of the parabola. The offset at the midpoint of the curve ($x = L/2$) is

$$y = ax^2 = [(g_2 - g_1)/2L](L/2)^2 = [(0.011 - 0.039)/2(129.36)](129.36/2)^2$$
$$= -0.45\text{ m}$$

6.6 d.

Station	Area, m²	Total Area	Average Area	Distance, m	Volume[a], m³
5 + 000	27.9				
		354	177.0	50	8850.0
5 + 050	326.1				
		1291.4	645.7	50	32,285.0
5 + 100	965.3				
				Total Volume	= 41,135.0

[a] Volume = distance × average area

6.7 a. The bases of the cross sections are constant (11 m). The heights of the cross sections vary between stations but can be determined by dividing the cross section areas by 11 m. The prismoidal formula is

$$V = (L/6)(A_1 + 4A_m + A_2)$$

Determine the volume between stations 1 + 130 and 1 + 200.

$$A_m = [(3.2 + 3.6)/2]11 = 37.4 \text{ m}^2$$
$$V = (70/6)[35 + 4(37.4) + 40] = 2620.3 \text{ m}^3$$

Determine the volume between stations 1 + 200 and 1 + 260.

$$A_m = [(3.6 + 4.1)/2]11 = 42.4 \text{ m}^2.$$
$$V = (60/6)[40 + 4(42.4) + 45 = 2546.0 \text{ m}^3]$$

Total volume = 2620.3 + 2546.0 = 5166.3 m³

CHAPTER 7

Water Purification and Treatment

Kenneth J. Williamson

OUTLINE

WATER DISTRIBUTION 83
Water Source ■ Transmission Line ■ The Distribution Network ■ Water Storage ■ Pumping Requirements

WATER QUALITY 85
Ion Balances ■ General Chemistry Concepts ■ Dissolved Oxygen Relationships

WATER TREATMENT 91
Sedimentation ■ Filtration ■ Softening ■ Chlorination ■ Fluoridation ■ Activated Carbon ■ Sludge Treatment

PROBLEMS 95

SOLUTIONS 97

WATER DISTRIBUTION

Water distribution systems involve a water source, a transmission line, a distribution network, pumping, and storage.

Water Source

The importance of the water source is primarily related to water elevation and water quality. The water elevation will determine the pumping required to maintain adequate flows in the network.

Water quality is largely determined by the source. Typical water sources include reservoirs, lakes, rivers, and underground aquifers. Surface water sources typically have low levels of dissolved solids but high levels of suspended solids. As a result, treatment of such sources requires coagulation, followed by sedimentation and filtration, to remove suspended solids. Groundwater sources are low in suspended solids but may be high in dissolved solids and reduced compounds. Groundwater, if high in dissolved solids, requires chemical precipitation or ion exchange for removal of dissolved ions.

Transmission Line

A transmission line is required to convey water from the sources and storage to the distribution system. Arterial main lines supply water to the various loops in the distribution system. The main lines are arranged in loops or in parallel to allow for repairs. Such lines are designed using the Hazen-Williams formula for single pipes (Equation 7.1) and the equivalent-pipe method for single loops.

$$v = kCr^{0.63}s^{0.54} \tag{7.1}$$

where
- v = pipe velocity (m/s)
- k = constant (2.79)
- C = roughness coefficient (100 for cast iron pipe)
- r = hydraulic radius (m)
- s = slope of hydraulic grade line (m/m)

The equivalent-pipe method involves two procedures: conversion of pipes of unequal diameters in series into a single length of pipe, and conversion of pipes in parallel flow from the same into a single pipe with a single diameter. The headloss is maintained through each conversion. Multiple loops are designed using Hardy Cross methodology.

The Distribution Network

The distribution network comprises the arterial mains, distribution mains, and smaller distribution piping. The controlling design variable for water distribution networks is the pressure under maximum flow, with a minimum value of 140 to 280 kPa and a maximum value of 690 kPa.

Flow for the water distribution network is calculated as the sum of domestic, irrigation, industrial and commercial, and fire requirements. The various flows for a given community are typically obtained from historical data of water use. Without such data, estimates can be made from average values listed in Table 7.1.

Table 7.1 Water flow rates

Flow	Method of Estimation
Average domestic flow	Population served, 0.4 m^3/cap-d, 3,000 to 10,000 persons/km^2
Irrigation flow	Maximum of 0.75 times average daily flow for arid climates
Industrial/commercial flow	Computed from known industries and area of commercial districts

Water Storage

Water storage is required to maintain pressure in the system, to minimize pumping costs, and to meet emergency demands. Typically the storage is placed so that the load center or distribution network is between the water source and the storage, as shown in Figure 7.1.

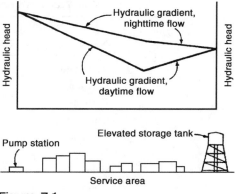

Figure 7.1

Storage requirements are calculated based upon variations in hourly flow and fire requirements. If the pumping capacity is set at some value less than the maximum hourly flow, then all flow requirements above that value will have to be provided by storage.

Pumping Requirements

The pumping system needs adequate capacity for design flows, taking into account the supply from storage. The required head must be adequate to maintain 140 kPa at the load center during maximum flows.

WATER QUALITY

The design of water treatment facilities requires calculations of concentrations and masses of various constituents in water and chemical additions. Common elements and ions of various constituents are listed in Table 7.2, and common water treatment chemicals are listed in Table 7.3. The equivalent weight is equal to the molecular weight divided by the absolute value of the valence.

Ion Balances

Electroneutrality requires that water have an equal number of equivalents of cations and anions. Often, bar graphs are used to show this relationship, as shown in Figure 7.2. From such diagrams, concentrations of alkalinity, carbonate (calcium and magnesium associated with alkalinity), and noncarbonate hardness (calcium and magnesium in other forms) can be easily identified.

General Chemistry Concepts

A number of concepts from general chemistry are required for engineering calculations related to water treatment.

Oxidation-Reduction Reactions

Oxidation-reduction (redox) reactions involve the transfer of electron from an electron donor to an electron acceptor. The easiest method to construct balanced redox reactions is through the use of half-reactions. A list of common half-reactions related to water treatment are in Table 7.4.

Table 7.2 Common elements and radicals

Name	Symbol	Atomic Weight	Valence	Equivalent Weight
Aluminum	Al	27.0	+3	9.0
Calcium	Ca	40.1	+2	20.0
Carbon	C	12.0	–4	
Chlorine	Cl	35.5	–1	35.5
Fluorine	F	19.0	–1	19.0
Hydrogen	H	1.0	+1	1.0
Iodine	I	126.9	–1	126.9
Iron	Fe	55.8	+2	27.9
			+3	
Magnesium	Mg	24.3	+2	12.15
			+4	
			+7	
Nitrogen	N	14.0	–3	
			+5	
Oxygen	O	6.0	–2	8.0
Potassium	K	39.1	+1	39.1
Sodium	Na	23.0	+1	23.0
Ammonium	NH_4^+	18	+1	18.0
Hydroxyl	OH^-	17.0	–1	17.0
Bicarbonate	HCO_3^-	61.0	–1	61.0
Carbonate	CO_3^{2-}	60.0	–2	30.0
Nitrate	NO_3^-	46.0	–1	46.0
Hypochlorite	OCl^-	51.5	–1	51.5

Henry's Law

Henry's Law states that the weight of any dissolved gas is proportional to the pressure of the gas.

$$C_{equil} = \alpha \, p_{gas} \tag{7.2}$$

where
  C_{equil} = equilibrium dissolved gas concentration
  P_{gas} = partial pressure of gas above liquid
  α = Henry's Law constant

Equilibrium Relationships

For an equilibrium chemical equation expressed as

$$A + B \rightleftharpoons C + D \tag{7.3}$$

the relationship of concentrations at equilibrium can be approximated as

$$K_{eq} = \frac{[C][D]}{[A][B]} \tag{7.4}$$

where
  K_{eq} = equilibrium constant
  [ ] = molar concentrations

Table 7.3 Common inorganic chemicals for water treatment

Name	Formula	Usage	Molecular Weight	Equivalent Weight
Activated carbon	C	Taste and odor	12.0	
Aluminum sulfate	$Al_2(SO_4)_3 \cdot 14.3H_2O$	Coagulation	600	100
Ammonia	NH_3	Chloramines, disinf.	17.0	
Ammonium fluosilicate	$(NH_4)_2SiF_6$	Fluoridation	178	
Calcium carbonate	$CaCO_3$	Corrosion control	132	66.1
Calcium fluoride	CaF_2	Fluoridation	78.1	
Calcium hydroxide	$Ca(OH)_2$	Softening	74.1	37.0
Calcium hypochlorite	$Ca(ClO)_2 \cdot 2H_2O$	Disinfection	179	
Calcium oxide	CaO	Softening	56.1	28.0
Carbon dioxide	CO_2	Recarbonation	44.0	22.0
Chlorine	Cl_2	Disinfection	71.0	
Chlorine dioxide	ClO_2	Taste and odor	67.0	
Ferric chloride	$FeCl_3$	Coagulation	162	54.1
Ferric hydroxide	$Fe(OH)_3$		107	35.6
Fluorosilicic acid	H_2SiF_6	Fluoridation	144	16.0
Oxygen	O_2	Aeration	32.0	
Sodium bicarbonate	$NaHCO_3$	pH adjustment	84.0	84.0
Sodium carbonate	Na_2HCO_3	Softening	106	53.0
Sodium hydroxide	$NaOH$	pH adjustment	40.0	40.0
Sodium hypochlorite	$NaClO$	Disinfection	74.4	
Sodium fluosilicate	Na_2SiF_6	Fluoridation	188	

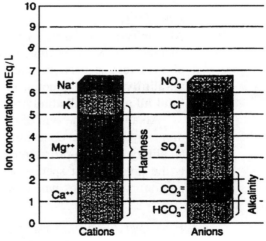

Figure 7.2

Some commonly used equilibrium constants are listed in Table 7.5.

A common equilibrium relationship is the disassociation of water, which is given as

$$H_2O \rightleftharpoons H^+ + OH^- \tag{7.5}$$

which has a K_{eq} value of 10^{-7}. The hydrogen ion concentration is typically represented by the pH value, or the negative log of the hydrogen ion concentration:

$$pH = -\log[H^+] \tag{7.6}$$

Table 7.4 Common half-reactions for water treatment

Reduced Element	Half-Reaction
Cl	$\frac{1}{2}Cl_2 + e^- \rightarrow Cl^-$
Cl	$\frac{1}{2}ClO^- + H^+ + e^- \rightarrow \frac{1}{2}Cl^- + \frac{1}{2}H_2O$
Cl	$\frac{1}{8}ClO_4^- + H^+ + e^- \rightarrow \frac{1}{8}Cl^- + \frac{1}{2}H_2O$
Fe	$\frac{1}{2}Fe^{2+} + e^- \rightarrow \frac{1}{2}Fe$
Fe	$Fe^{3+} + e^- \rightarrow Fe^{2+}$
Fe	$\frac{1}{3}Fe^{3+} + e^- \rightarrow \frac{1}{3}Fe$
I	$\frac{1}{2}I_2 + e^- \rightarrow I^-$
N	$\frac{1}{8}NO_3^- + 5/4 H^+ + e^- \rightarrow \frac{1}{8}NH_4^+ + \frac{3}{8}H_2O$
N	$\frac{1}{5}NO_3^- + \frac{6}{5}H^+ + e^- \rightarrow \frac{1}{10}N_2 + \frac{3}{5}H_2O$
O	$\frac{1}{4}O_2 + H^+ + e^- \rightarrow \frac{1}{2}H_2O$

Table 7.5 Common equilibrium constants

Equation	K_{eq}
$H_2CO_3 \rightleftharpoons H^+ + HCO_3^-$	$10^{-6.4}$
$HCO_3^- \rightleftharpoons H^+ + CO_3^{2-}$	$10^{-10.3}$
$NH_3 + H_2O \rightleftharpoons NH_4^+ + OH^-$	$10^{-4.7}$
$CaOH^+ \rightleftharpoons Ca^{2+} + OH^-$	$10^{-1.5}$
$MgOH^+ \rightleftharpoons Mg^{2+} + OH^-$	$10^{-2.6}$
$HOCl \rightleftharpoons H^+ + OCl^-$	$10^{-7.5}$

Alkalinity

Alkalinity is a measure of the ability of water to consume a strong acid. In the alkalinity test, water is titrated with a strong acid to pH 6.4 and to pH 4.5. The amount of acid that is used to reach the first pH endpoint is termed the **carbonate alkalinity**, and the amount of acid needed to reach the second end point is the **total alkalinity**. Alkalinity is expressed as calcium carbonate.

Alkalinity is the sum of bicarbonate, carbonate, and hydroxide concentrations minus hydrogen ion concentration. For most sources to be used for drinking water, alkalinity can be approximated as carbonate and bicarbonate. Carbonate alkalinity is the predominant species when the pH is above 10.8, and bicarbonate is the predominant species when the pH is between 6.9 and 9.8.

Solubility Relationships

The equilibrium between a compound in its solid crystalline state and its ionic form in solution, where

$$X_a Y_b = aX^{b+} + bY^{a-} \tag{7.7}$$

is given by

$$K_{sp} = [X^{b+}]^a [Y^{a-}]^b \tag{7.8}$$

Table 7.6 Common solubility products

Compound	K_{sp}
Magnesium carbonate	4×10^{-5}
Magnesium hydroxide	9×10^{-12}
Calcium carbonate	5×10^{-9}
Calcium hydroxide	8×10^{-6}
Aluminum hydroxide	1×10^{-32}
Ferric hydroxide	6×10^{-38}
Ferrous hydroxide	5×10^{-15}
Calcium fluoride	3×10^{-11}

where K_{sp} = solubility product. Solubility products for common precipitation reactions in water treatment are given in Table 7.6.

Dissolved Oxygen Relationships

The dissolved oxygen concentration in a stream is determined by the rate of oxygen consumption, commonly caused by the discharge of oxygen-demanding substances, and the rate of reoxygenation from the atmosphere.

Biochemical Oxygen Demand

The concentration of oxygen-demanding substances in a river is commonly expressed as biochemical oxygen demand or BOD as described in Chapter 5. BOD is measured as the oxygen that is removed in a 20° C, five-day test and is expressed as BOD_5. This value has to be converted to an ultimate BOD or L value at the temperature of interest as

$$L = \frac{BOD_5}{1 - e^{-5k_{20}}} \quad (7.13)$$

where k_{20} = BOD decay constant at 20° C. Typical k_{20} values for organic wastes range from about 0.20 to 0.30.

The rate of oxygen consumption by BOD is related to the ultimate BOD concentration as

$$r_{BOD} = k_T L \quad (7.14)$$

Oxygen Deficit

The dissolved oxygen in water is determined by Henry's Law, based upon a 20 percent partial pressure in the atmosphere. Saturation values as a function of temperature can be extrapolated from known values of 14.6, 11.3, 9.1, and 7.5 mg/L at 0, 10, 20, and 30° C, respectively.

At a given temperature the dissolved oxygen concentration can be expressed also as a deficit or

$$D = (C_{sat} - C) \quad (7.15)$$

where
C_{sat}, C = saturation and actual dissolved oxygen concentration (mg/L)
D = dissolved oxygen deficit at given temperature

Mixing

Assuming complete mixing in a river, the concentration of dissolved oxygen and ultimate BOD in a receiving stream is related as

$$C_0 = \frac{Q_r C_r + Q_w C_w}{Q_r + Q_w} \quad (7.16)$$

where

C_r, C_w = concentration of constituents in river and waste, respectively
Q_r, Q_w = flow rate of river and waste, respectively

Reoxygenation

Reoxygenation occurs by diffusion of oxygen from the atmosphere to the river. The rate of reoxygenation or reaeration is given as

$$r_{\text{reoxy}} = k_2 D \quad (7.17)$$

where k_2 = reaeration coefficient (1/d).

Oxygen Sag Model

The dissolved oxygen concentration in a stream after the addition of oxygen-demanding wastes can be expressed in differential form as

$$\frac{dD}{dt_r} = r_{\text{BOD}} - r_{\text{reoxy}} \quad (7.18)$$

where t_r = travel time in the river from the point of waste addition (d).

The resulting dissolved oxygen profile is shown in Figure 7.3. After addition of the waste, the dissolved oxygen decreases to a maximum deficit (D_c) or minimum dissolved oxygen level at a distance x_c or travel t_c. Past this point, the deficit decreases as the river recovers.

Equation (7.18) can be integrated to

$$D = \frac{kL_0}{k_2 - k}\left(e^{-kt_r} - e^{-k_2 t_r}\right) + D_0 e^{-k_2 t_r} \quad (7.19)$$

where D_0, L_0 = oxygen deficit and ultimate BOD concentration at point of waste discharge in the river, respectively.

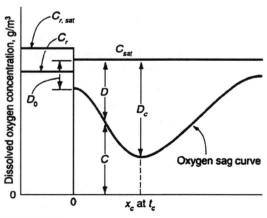

Figure 7.3

WATER TREATMENT

Water treatment is used to alter the quality of water to make it chemically and bacteriologically safe for human consumption. Common sources are groundwater and surface waters. Groundwater treatment may involve removal of pathogens, removal of iron and manganese, and removal of hardness (Figure 7.4). Surface water treatment typically involves simultaneous removal of pathogens, suspended solids, and taste- and odor-causing compounds (Figure 7.5). All of these treatment processes are composed of unit processes.

Sedimentation

After flocculation, the larger particles are removed by sedimentation. Settling is also used in water treatment after oxidation of iron or manganese. Flocculated particles have densities of about 1400 to 2000 kg/m.

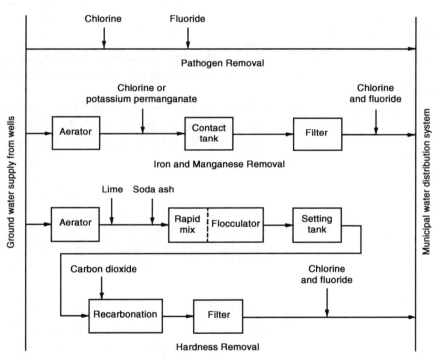

Figure 7.4

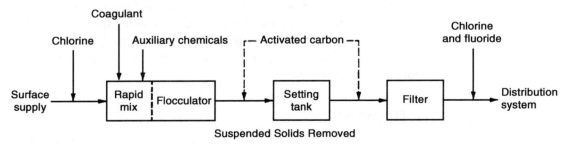

Figure 7.5

Settling velocities of particles are determined by Stokes' law for Reynolds numbers less than 0.3:

$$v_s = \frac{g(\rho_p - \rho_w)d_p^2}{18\mu} \quad (7.20)$$

where
v_s = terminal settling velocity (m/s)
g = gravitational constant (9.8 m/s^2)
d_p = particle diameter (m)
μ = water viscosity (0.001 kg/m-s)

For particles coming into a sedimentation basin, the particles enter at all depths. A critical settling velocity related to the sedimentation basin can be calculated as

$$v_{sc} = \frac{Q}{A_s} \quad (7.21)$$

where
Q = flow rate into sedimentation basin (m^3/s)
A_s = surface area of sedimentation basin (m^2)

In standard sedimentation theory, if the settling rate of the particles to be removed is greater than v_{sc}, then the particles will be 100% removed. For particles with v_s less than v_{sc}, the particles will be only partially removed, with the removal given as

$$R = \frac{v_s}{v_{sc}} \quad (7.22)$$

where R = decimal removal.

The critical settling velocities used for design are typically expressed as an overflow rate.

Filtration

Filtration in water treatment is commonly accomplished with granular filters. Media for such filters may include sand, charcoal, and garnet; the filters can involve one, two, or several different filter media. The water is applied to the top of the filter and is collected through underdrains.

Filtration is a complex process involving entrapment, straining, and absorption. As the filtration process proceeds, the headloss associated with the water flow through the porous media increases as the filtered material accumulates in the pores. In addition, the number of particles passing through the filter increases, with a subsequent increase in effluent turbidity. When either the allowable headloss or allowable effluent quality is exceeded, then the filter has to be backwashed to remove the accumulated material.

Softening

The removal of hardness is accomplished using line/soda ash softening or ion exchange.

Lime/Soda Ash Softening

Lime/soda ash softening occurs by the addition of calcium hydroxide and sodium bicarbonate to form a chemical precipitate, which is removed by sedimentation and filtration.

For waters containing excess, the calcium hydroxide reacts with carbon dioxide and carbonate hardness as

$$CO_2 + Ca(OH)_2 \rightarrow CaCO_3 + H_2O$$
$$Ca(HCO_3)_2 + Ca(OH)_2 \rightarrow CaCO_3 + 2H_2O$$
$$Mg(HCO_3)_2 + 2Ca(OH)_2 \rightarrow 2CaCO_3 + Mg(OH)_2 + 2H_2O$$

The sodium bicarbonate and calcium hydroxide react with the noncarbonate hardness as

$$Ca^{2+} + Na_2CO_3 \rightarrow CaCO_3 + H_2O$$
$$Mg^{2+} + Ca(OH)_2 \rightarrow Mg(OH)_2 + Ca^{2+}$$

Recarbonation is required after treatment to remove the excess lime, magnesium hydroxide, and carbonate, reducing the pH to about 8.5 to 9.5.

Based upon the foregoing equations, the requirements of lime (L) and soda ash (SA) in mEq/L are

$$L = CO_2 + HCO_3^- + Mg^{2+} \qquad (7.23)$$

$$SA = Ca^{2+} + Mg^{2+} - Alk \qquad (7.24)$$

The lime requirement needs to be increased about 1 Eq/m^3 to raise the pH to allow precipitation of the magnesium hydroxide.

Ion Exchange

Ion exchange processes are used to remove ions of calcium, magnesium, iron, and ammonium. The exchange material is a solid that has functional groups that replace ions in solution for ions on the exchange material. Ion exchange materials can be made to remove cations or anions. Most materials are regenerated by flushing with solutions with high concentrations of either H^+ or Na^+.

Ion exchange process is controlled by the exchange capacity and the selectivity. The **exchange capacity** expresses the equivalents of cations or anions that can be exchanged per unit mass. The **selectivity** for ion B as compared to ion A is expressed as

$$K_{B/A} = \frac{\chi_{R-A}[A^+]}{\chi_{R-B}[B^+]} \qquad (7.25)$$

where

$K_{B/A}$ = selectivity coefficient for replacing A on resin with B;
χ_{R-A}, χ_{R-B} = mole fractions of A and B for the absorbed species.

Chlorination

Chlorination is the most common form of disinfection for treatment of water. Chlorine is a strong oxidation agent that destroys organisms by chemical attack. Chlorine is usually added in the form of chlorine gas at a level of 2 to 5 mg/L. Other common forms are chlorine dioxide, sodium hypochlorite, and calcium hypochlorite.

Chlorine in water undergoes an equilibrium reaction to form HOCl and OCl⁻. The hypochlorite is a weak acid as listed in Table 7.5. HOCl is a more effective disinfectant than OCl⁻; as such, the effectiveness is strongly dependent upon pH.

The effectiveness of chlorination depends upon the time of contact, the chlorine concentration, and the concentration of organisms. The effect of time is modeled as

$$\frac{N_t}{N_0} = e^{-kt^m} \quad (7.26)$$

where
N_t, N_0 = number of organisms at time t and time zero
k = decay coefficient (1/d)
t = time (d)
m = empirical coefficient (usually 1)

Fluoridation

Fluoridation of water supplies has been shown to reduce dental caries dramatically. Optimum concentrations are about 1 mg F/L. The most commonly used compounds are sodium fluoride, sodium fluosilicate, and fluorosilicic acid. Fluoride is usually added after coagulation or lime/soda ash softening, because high calcium concentration can result in the fluorides precipitating.

Activated Carbon

Activated carbon is added in water treatment to adsorb taste- and odor-causing compounds and to remove color. The process usually involves the direct addition of powdered activated carbon, with removal in the sedimentation and filtration units. The required dosage is usually determined empirically using laboratory tests.

Sludge Treatment

The sources of sludge in water treatment include sand, silt, chemical sludges, and backwash solids. These sludge are typically concentrated in settling basins or lagoons. The sludges can be further concentrated by drying on sand drying beds or by centrifugation.

PROBLEMS

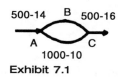
Exhibit 7.1

7.1 Find a single pipe to replace the pipe loop shown in Exhibit 7.1.

7.2 How many grams of oxygen are required to burn 1 gram of methane?
$$CH_4 + O_2 \rightarrow CO_2 + H_2O$$

7.3 A liter of water is at equilibrium with an atmosphere containing a partial pressure of 0.1 atm of CO_2. How many grams are dissolved in the water? ($\alpha = 2.0$ g/L-atm)

7.4 How much HOCl is present in a solution containing 0.1M chlorine at pH 8?

7.5 How many grams of fluoride would be present in a solution saturated with calcium fluoride?

7.6 A 5-d BOD and ultimate BOD are measured at 180 mg/L and 200 mg/L, respectively. What is the decay coefficient?

7.7 A wastewater with a dissolved oxygen concentration of 1 mg/L is discharged to a river. The river is at 20° C and saturated with dissolved oxygen. If the flows are 2.8×10^{-2} m/s and 2.8 m/s for the wastewater and the river, respectively, what is the oxygen deficit after mixing?

7.8 A water treatment plant is required to produce an average treated water flow 14.4×10^4 m^3/d from a water source of the following characteristics:

Chlorines	92 mg/L
Potassium	31 mg/L
Sodium	14 mg/L
Sulfates	134 mg/L
Calcium	94 mg/L
Magnesium	28 mg/L
Alkalinity	135 mg/L as $CaCO_3$
pH	7.8
Temperature	21° C
Total solids	720 mg/L

Determine an ion balance for the water.

7.9 A water contains silt particles with a uniform diameter of 0.02 mm and a specific gravity of 2.6. What removal is expected in a clarifier with an overflow rate of 12 m/d (300 gal/ft^2-d))?

7.10 The carbonaceous oxygen demand of 200 mg/L of glycine (H_2NCH_2COOH) is
a. less than the nitrogenous oxygen demand
b. depends on the reaction rate
c. greater than 120 mg/L
d. greater than 100 mg/L

7.11 A completely mixed lagoon has a waste flow of 1 m³/s of a BOD waste with a decay coefficient of 0.3/d and a concentration of 100 mg/L. If the effluent BOD must be 20 mg/L or less, then
a. the volume required is about 1×10^6 m³
b. the volume required is about 2×10^6 m³
c. the volume required is about 0.5×10^6 m³
d. an increase in temperature would increase the volume required

7.12 A completely mixed reactor with cell recycle is designed to treat a municipal waste. Assuming removal kinetics follow the equation:

$$r_{su} = -\frac{k \times S}{(K_s + S)}$$

which of the following statements is correct?
a. The effluent substrate concentration decreases with an increase in θ_c.
b. The food:microorganism level is independent of θ_c.
c. Microbe concentrations will be smaller than in the no-recycle case.
d. θ_c is independent of effluent quality.

7.13 Which of these statements are correct?
1. Chlorination of wastewater effluents requires more chlorine than chlorination of drinking water.
2. Chlorination of wastewater effluents requires three moles of chlorine for each mole of ammonia.
3. Chlorination of wastewater effluents is used to improve effluent quality.
4. Chlorination of wastewater effluents oxidizes other chemicals such as ferrous iron.

a. All of the above statements are correct.
b. None of the above statements is correct.
c. Only statements 1 and 3 are correct.
d. Only statements 1, 3, and 4 are correct.

7.14 A chemical is removed in a completely mixed reactor at a zero-order rate of 0.25 mg/L-d. The influent concentration is 10 mg/L and the hydraulic detention time is 10 days. For this reactor, which of these statements are correct?
1. The effluent concentration would be the same as that for a batch reactor with a 2-day retention time.
2. The effluent concentration will be about 7.5 mg/L.
3. The effluent quality would improve with longer detention times.
4. The effluent quality would improve with increased mixing.

a. All of the above statements are correct.
b. None of the above statements is correct.
c. Only statements 1 and 2 are correct.
d. Only statements 1, 3, and 4 are correct.

SOLUTIONS

7.1 Assume a flow in A-B-C of 4 cfs. Use the nomograph in Exhibit 7.1a.

$$h_L \text{ in AB} = 5.8'/1000'$$
$$= 2.9 \text{ ft}$$

$$h_L \text{ in BC} = 3.0'/1000'$$
$$= 1.5'$$

$$\text{Total } h_L = 4.5'$$
$$= 4.5'/1000'$$

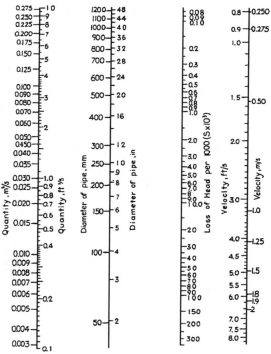

Exhibit 7.1a Flow in old cast iron pipes. (Hazen-Williams $C = 100$)

Use $\phi = 14.5"$, $L = 1000'$ for a h_L of 4.5' for A-C. then $Q = 1.4$ cfs.

$$Q_{\text{Total}} = 4 + 1.4 = 5.4 \text{ cfs}$$
$$\text{Equivalent diameter for loop} = 16.2"$$
$$\text{Equivalent length for loop} = 1000'$$

7.2 Balance the equation as

$$CH_4 + 2O_2 \rightarrow CO_2 + 2H_2O$$
$$gO_2 = 1 \text{ g } CH_4 \times \frac{1 \text{ mol } CH_4}{16 \text{ g } CH_4} \times \frac{2 \text{ moles } O_2}{1 \text{ mole } CH_4} \times \frac{32 \text{ g } O_2}{1 \text{ mole } O_2}$$
$$= 4 \text{ g } O_2$$

7.3

$$C_{equil} = \alpha P \text{ gas}$$
$$= 2.0 \text{ g}/1-\text{atm} \times 0.1 \text{ atm}$$
$$= 2.0 \text{ g/liter}$$

$$M_{CO_2} = C \times V$$
$$= 0.2 \text{ g}/1 \times 1 \text{ liter}$$
$$= 0.2 \text{ g}$$

7.4

$$[H^+] = 10^{-pH} = 10^{-8}$$
$$HOCl \Leftrightarrow H^+ + OCl^-$$
$$\frac{[H^+][OCl^-]}{[HOCl]} = 10^{-7.5}$$
$$\frac{[H^+](0.1-HOCl)}{[HOCl]} = 10^{-7.5}$$
$$\frac{(0.1-[HOCl])}{[HOCl]} = 3.16$$
$$[HOCl] = 0.024 \text{ M}$$

7.5

$$CaF_2 \rightarrow Ca^{2+} + 2F^-$$
$$[Ca^{2+}][F^-]^2 = 3 \times 10^{-11}$$
$$\left[\frac{1}{2}F^-\right][F^-]^2 = 3 \times 10^{-11}$$
$$[F^-] = [2(2 \times 10^{-11})]^{1/3}$$
$$= 4.2 \times 10^{-4} \text{ M}$$
$$= 8.0 \times 10^{-3} \text{ g/liter}$$

7.6

$$1 - e^{-k(5d)} = BOD_5/L$$
$$= 180/300 = 0.6$$
$$k = 0.18/d$$

7.7

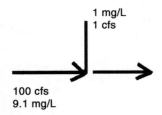

Exhibit 7.7

$$C = \frac{(100 \text{ cfs})(9.1 \text{ mg/L}) + (1 \text{ cfs})(1 \text{ mg/L})}{101 \text{ cfs}}$$
$$= 9.0 \text{ mg/L}$$
$$D = C_{sat} - C$$
$$= 0.1 \text{ mg/L}$$

7.8 The milliequivalents per liter for the ions are:

Ion	Conc (mg/L)	Equiv. wt.	Conc (mEq/L)
Na^+	14	23	0.61
K^+	31	39	0.79
Mg^{2+}	28	12.2	2.30
Ca^{2+}	94	20	4.7
		Total	8.4
SO_4^{2-}	134	48	2.79
Cl^-	92	35.5	2.59
$HCO_3^- + CO_3^{2-}$	135	50	2.70
		Total	8.1

When the concentration of the alkalinity (expressed as $CaCO_3$) is converted to mEq/L, it will be equal to the sum of the concentrations of bicarbonate and carbonate.

7.9

$$v_s = \frac{g(\rho_p - \rho_w)d_p^2}{18\mu}$$

$$= \frac{(9.8 \text{ m/s}^2)(1600 \text{ kg/m}^3)(2\times 10^{-5} \text{ m})^2}{(18)(1\times 10^{-3} \text{ N-s/m}^2)}$$

$$= 3.5\times 10^{-5} \text{ m/s}$$

$$v_{sc} = \frac{300 \text{ gal}}{\text{ft}^2\text{-d}} \times \frac{0.0017 \text{ m/hr}}{1 \text{ gal/ft}^2\text{-d}} \times \frac{1 \text{ hr}}{3600 \text{ s}}$$

$$v_{sc} = 12 \text{ m/d} \times \frac{d}{86{,}400 \text{ s}} = 1.4\times 10^{-4} \text{ m/s}$$

$$= 1.4\times 10^{-4} \text{ m/s}$$

$$k = \frac{3.5\times 10^{-5} \text{ m/s}}{14\times 10^{-5} \text{ m/s}} = 0.25 \text{ or } 25\%$$

7.10 c.

7.11 a.

$$\frac{A}{A_0} = \frac{1}{1 + K_1 \frac{v}{Q}}$$

$$0.2 = \frac{1}{1 + K_1 \frac{v}{Q}}$$

$$1 + K_1 \frac{v}{Q} = 5$$

$$v = \frac{4Q}{K_1} = \frac{3.46 \times 10^5}{0.3} = 1.1 \times 10^6 \text{ m}^3$$

7.12 a.

7.13 d.

7.14 c.

CHAPTER 8

Surveying

Indranil Goswami

OUTLINE

GLOSSARY OF SURVEYING TERMS   102

BASIC TRIGONOMETRY   103

TYPES OF SURVEYS   103

COORDINATE SYSTEMS   103
State Plane Coordinate System (SPCS) ■ Global Positioning Systems (GPS)

STATIONING   104

CHAINING TECHNIQUES   104
Tension Correction ■ Temperature Correction ■ Sag Correction

DIFFERENTIAL LEVELING   105

ANGLES AND DISTANCES   106
Azimuth ■ Latitude and Departure ■ Northings and Eastings ■ Interior and Exterior Angles

TRAVERSE CLOSURE   109
Compass Rule ■ Transit Rule ■ Application of the Transit Rule

AREA OF A TRAVERSE   111
By Coordinates ■ By Double Meridian Distance

AREA UNDER AN IRREGULAR CURVE   113

PROBLEMS   115

SOLUTIONS   117

Approximately 11% of the civil FE/EIT exam concerns surveying topics; candidates can expect to see roughly seven surveying questions. According to NCEES, the following specific topics may be covered:

■ Angles, distances, and trigonometry

■ Area computations

■ Closure

- Coordinate systems (e.g., GPS, state plane)
- Curves (vertical and horizontal)
- Earthwork and volume computations
- Leveling (e.g., differential, elevations, and percent grades)

This chapter covers key terms, concepts, and analytical techniques of surveying that you may encounter on the exam. We begin with a glossary of terms and a brief summary of basic trigonometry; this information will be useful throughout the rest of the chapter. Coverage of vertical and horizontal curves can be found in Chapter 6, "Transportation Engineering." That chapter also contains some additional discussion of earthwork and volume computations.

GLOSSARY OF SURVEYING TERMS

Backsight (BS) Elevation (rod reading) obtained by sighting back station. Also called Plus Sight (+S).

Benchmark A fixed reference point or object, the elevation of which is known.

Contour An imaginary line of constant elevation on the ground surface.

Deflection angle An angle measured to a line from the extension of the preceding line.

Departure The orthographic projection of a line on the east-west axis of the survey. East departures are considered positive.

Foresight (FS) Elevation (rod reading) obtained by sighting forward station. Also called Minus Sight (– S).

Height of instrument (HI) The elevation of the line of sight of the telescope above the survey station or control point.

Height of target The elevation of the target or prism above the survey station or control point.

Least count The smallest graduation shown on a vernier. The least count of an instrument is the smallest possible measurement that can be made without interpolation.

Level surface A curved surface, every element of which is normal to a plumb line.

Latitude (traverse) The orthographic projection of a line on the north-south axis of the survey. North latitudes are considered positive.

Latitude (astronomical) Angle measured along a meridian north (positive) and south (negative) from the equator. Latitude varies from 0° to 90°.

Longitude Angle measured at the pole, east or west from the prime meridian. Longitude varies from 0° to 180° east or 180° west.

Meridian (astronomical) An imaginary line on the earth's surface having the same astronomical longitude at every point.

Traverse A succession of lines for which distance and horizontal angles are measured. A traverse may be closed, where the traverse ends at the point from which it started, or open.

Vernier An auxiliary scale placed alongside the main scale of an instrument, by means of which the fractional parts of the least division of the main scale can be measured precisely.

Zenith The zenith is the point on the celestial sphere where the gravity vector, extended upward, intersects it.

Zenith angle An angle formed between two intersecting lines in a vertical plane where one of these lines is directed toward the zenith.

BASIC TRIGONOMETRY

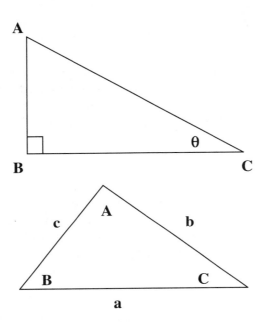

For a Right Triangle

$$\sin\theta = \frac{AB}{AC}$$

$$\cos\theta = \frac{BC}{AC}$$

$$\tan\theta = \frac{AB}{BC}$$

Pythagorean theorem $AB^2 + BC^2 = AC^2$

For Any Triangle

Law of Sines

$$\frac{a}{\sin A} = \frac{b}{\sin B} = \frac{c}{\sin C}$$

Law of Cosines

$$a^2 = b^2 + c^2 - 2bc\cos A$$
$$b^2 = a^2 + c^2 - 2ac\cos B$$
$$c^2 = a^2 + b^2 - 2ab\cos C$$

TYPES OF SURVEYS

Plane surveying methods consider the surface of the earth as a plane. Curvature of the earth surface is neglected. This type of survey is appropriate for small geographical areas.

Geodetic surveying takes into account the true (near-spherical or spheroidal) shape of the earth. When the survey covers a large geographical area, neglecting the curvature of the earth introduces significant errors.

COORDINATE SYSTEMS

State Plane Coordinate System (SPCS)

Standardized in 1983 (*NOAA Manual NOS NGS 5*), the state plane coordinate system is a map projection system based on the North American Datum of 1983 (NAD 83). Given the geodetic coordinates (latitude and longitude) of a point, the equations in SPCS can be used to convert these coordinates to state plane coordinates (northings and eastings).

Global Positioning System (GPS)

The global positioning system is based on a constellation of nongeosynchronous satellites, whose position at any time is known to a very high degree of precision. The GPS system was developed by the U.S. Department of Defense in 1978 and is officially named NAVSTAR. GPS receivers on the ground calculate the distance to these known reference points by measuring a time lag (phase shift) between a coded signal generated by the satellite and an identical signal generated by the receiver. When distances from four satellites are measured simultaneously, the intersection of the four imaginary spheres reveals the location of the receiver. Knowledge of the elevation of the point on the ground can reduce the number of necessary satellites to three. GPS data are susceptible to errors due to atmospheric and ionospheric effects, ephemeris and atomic clock errors, numerical errors, and errors due to multipath effects. Moving receivers, such as those in vehicles, are less susceptible to multipath errors.

Differential GPS may be used to achieve subcentimeter accuracy. The sum total of all errors inherent in measurements is estimated using a second receiver located at a known position. This error estimate is then used to adjust the computed position of other receivers in the same general locale.

STATIONING

The concept of stations is commonly used for linear projects, such as in transportation engineering. In the United States, a common convention is to utilize 100-foot stations. Thus, a distance of 320.76 feet may be expressed as 3 +20.76 stations. Earthwork (haulage) costs are often expressed in typical units of $/yd^3 - sta. This means the cost of hauling a soil volume of 1 yd^3 a distance of 1 station (100 ft).

CHAINING TECHNIQUES

Distances used in surveying are generally horizontal distances, not sloped distances (e.g., property lines on deeds, stationing, layout of building corners, etc.). Even though chaining as a distance measurement method has been rendered obsolete by modern optical instruments, the history of surveying has involved the use of chains and tapes (typically steel) for measuring distance. The results obtained by chaining need to be corrected for extension errors (created by the application of a tension on the tape to prevent excessive sag), thermal elongation errors (created by the difference between the average ambient temperature and the tape's standardization temperature), and sag errors (created by the self-weight induced sag of a tape that is held between two elevated supports).

Tension Correction

$$e_P = \frac{(P - P_0)L}{AE} \quad (8.1)$$

where
  P = applied tension
  P_0 = tension at which tape or chain was standardized
  L = length of tape or chain
  A = cross-sectional area of tape or chain
  E = modulus of elasticity of tape or chain (steel has E = 29,000 kips/in^2 or 200 GPa)

Note: Applied tension greater than the standard tension makes the tape longer than its standard length, thereby recording a lesser tape reading (negative error), and the correction should be added (positive correction).

Temperature Correction

$$e_T = L\alpha(T - T_0) \tag{8.2}$$

where
- α = coefficient of thermal expansion of chain or tape (steel has $\alpha = 6.5 \times 10^{-6}/°F$ = $11.6 \times 10^{-6}/°C$)
- L = measured length
- T = ambient temperature
- T_0 = standardization temperature

Note: Ambient temperature greater than the standard temperature makes the tape longer than its standard length, thereby recording a lesser tape reading (negative error), and the correction should be added (positive correction).

Sag Correction

$$e_S = \frac{w^2 L^3}{24 P^2} \tag{8.3}$$

where
- w = weight of tape per unit length
- L = horizontal distance between level supports
- P = applied tension

Note: Tape sag makes recorded length greater than actual (positive error), and the correction should be subtracted (negative correction).

DIFFERENTIAL LEVELING

Differential leveling is the process of measuring the difference in elevation between points as illustrated in Figure 8.1. A transit level is used in combination

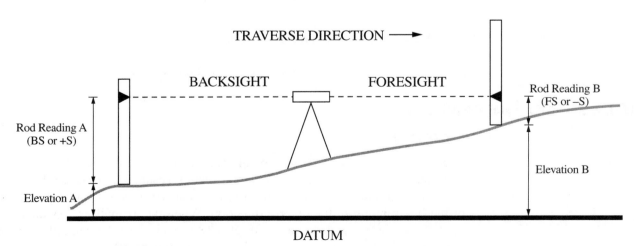

Figure 8.1 Differential leveling

Table 8.1 Differential leveling example

Station	BS	HI	FS	Elevation	Notes
BM-A	6.35			1000.00	Benchmark – marker no. ES-12
		1006.35			
B	9.29		2.45	1003.90	
		1013.19			
C	6.15		3.78	1009.41	
		1015.56			
BM-D			5.56	1010.00	Benchmark – marker no. ES-13
Σ =	**21.79**	Σ =	**11.79**		**Check** ΣBS − ΣFS = 21.79 −11.79 = +10.00 Ending elevation = 1000.00 + 10.00 = 1010.00 OK

with a rod or staff to determine elevations of points by obtaining differential vertical measurements. All elevations are referenced to a known elevation or **benchmark**. Table 8.1 shows typical field data for a series of staff readings obtained at successive stations A, B, C, ... in a traverse. In this example, the traverse is started at station A (known elevation 1000 ft) and ended at station D (elevation 1010 ft). In each case, the height of instrument (HI) is equal to the elevation of the back station plus the backsight (BS) (e.g., 1000.00 + 6.35 = 1006.35) or the elevation of the forward station plus the foresight (FS) (e.g., 1003.90 + 2.45 = 1006.35).

As the telescope is leveled, the line of sight is horizontal (or level). A graduated rod is held vertically at a point of known elevation. The elevation of the benchmark *plus* the rod reading at the benchmark gives the HI, which is the elevation of the line of sight. If the rod is now held vertically at some other location, its unknown elevation can be calculated as the elevation of the line of sight *minus* the rod reading. Thus, with successive locations (1, 2, . . .) of the level, we may write as follows:

$$HI_1 = Elevation_A + BS_A$$
$$Elevation_B = HI_1 - FS_B$$
$$HI_2 = Elevation_B + BS_B$$
$$Elevation_C = HI_2 - FS_C$$

etc.

which can also be expressed as follows:

$$Elevation_A + BS_A = Elevation_B + FS_B$$

ANGLES AND DISTANCES

Azimuth

The azimuth of a line is the horizontal angle measured to the line from a specific meridian (usually north) in a particular direction (usually clockwise). Figure 8.2 shows the azimuths of lines AB and BA. The back azimuth of a line is the azimuth of the line running in the reverse direction. When the azimuth is less than 180°, the back azimuth equals the azimuth plus 180°, and when the azimuth is greater than 180°, the back azimuth equals the azimuth minus 180°.

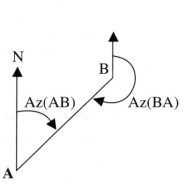
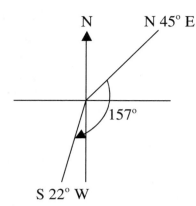

Figure 8.2 Azimuth and back-azimuth of line bearings

Figure 8.3 Bearings of lines and angle between lines

Bearings of lines are directional (horizontal) angles with respect to a meridian (north or south), measured at the originating point on the line. For example, the bearing of a line headed in the north-east direction can be written as N45°E.

The angle between two lines whose bearings are given may be calculated as the difference between their azimuths. For example, if the two lines are N45°E and S22°W, then the first step will be to describe them as azimuths. It is helpful to visualize the angles: S22°W is in the third quadrant; 22° west of south, as shown in Figure 8.3. The azimuth of N45°E is 45°, and the azimuth of S22°W is 22° + 180° = 202°. Therefore, the angle between the two lines is 202° − 45° = 157°.

Latitude and Departure

In the rectangular coordinate system, where the north-south meridian serves as the *y*-axis and the east-west line serves as the *x*-axis, the projections of a line are termed **latitude** (*y*-projection) and **departure** (*x*-projection), as shown in Figure 8.4. The departure is considered positive to the east (i.e., the line shows an *increase in easting*), and the latitude is considered positive to the north (i.e., the line shows an *increase in northing*).

If the bearing of the line shown in Figure 8.4 is NθE, where θ is the angle between the north meridian and the line (also known as the **azimuthal angle**), then the latitude is given by $L\cos\theta$ and the departure by $L\sin\theta$, where L is the length of line AB.

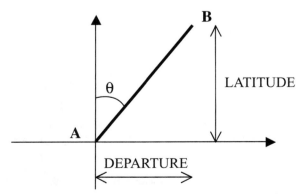

Figure 8.4 Latitude and departure

Northings and Eastings

Whereas latitudes and departures are north-south and east-west projections, respectively, of a line segment, **northings** and **eastings** are coordinates of points in a traverse. Thus, for line segment AB, the latitude is the difference of the northings of points A and B, and the departure is the difference of the eastings.

Interior and Exterior Angles

For a closed traverse (polygon with n sides), the following relations must hold:

$$\text{Sum of all exterior angles} = 180° \, (n + 2)$$
$$\text{Sum of all interior angles} = 180° \, (n - 2)$$

For successive lines in a traverse, interior angles may be calculated from azimuths:

$$\text{Azimuth}_1 + 180° - \text{Interior angle} = \text{Azimuth}_2$$

This may also be stated as:

$$\text{Interior angle} = \text{Azimuth}_1 + 180° - \text{Azimuth}_2 = \text{Back azimuth}_1 - \text{Forward azimuth}_2$$

The bearings of lines AB, BC, CD, and DA in Figure 8.5 may be written as follows:

AB	S 60° E
BC	S 17° E
CD	S 75° W
DA	N 23° W

In each of these cases, either the north or south direction is used as the reference meridian, utilizing the acute angle. This notation is traditional. In order to calculate interior angles, one should first convert each of these bearings to an

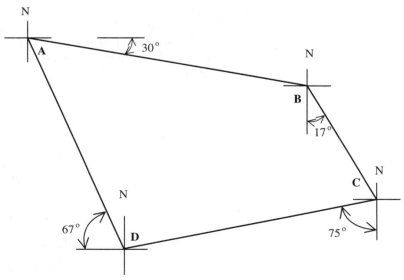

Figure 8.5 Illustration of relation between bearings and interior angles

equivalent azimuth (from north), with the angle measured clockwise, that is, toward the east. Thus, the azimuths are:

$$Az_{AB} = 120°, Az_{BC} = 163°, Az_{CD} = 255°, Az_{DA} = 337°$$

Using these azimuths, we can calculate the interior angles at A, B, C, and D as:

Angle A = Back azimuth DA − Forward azimuth AB
= 337° + 180° − 120° = 397° = 37° (modulo 360°)

Angle B = Back azimuth AB − Forward azimuth BC
= 120° + 180° − 163° = 137°

Angle C = Back azimuth BC − Forward azimuth CD
= 163° + 180° − 255° + 180° = 88°

Angle D = Back azimuth CD − Forward azimuth DA
= 255° + 180° − 337° + 180° = 98°

Note that the sum of the interior angles is 360° (as appropriate for a quadrilateral).

TRAVERSE CLOSURE

A closed traverse must, in a perfect world, be closed; that is, it must have no closure error. However, in starting a traverse from a particular station and following several paths to finally return to the originating station, there are many length and angle measurements, each of which could have some degree of error. Errors can be instrument error or human error. The purpose of traverse closure methods is to distribute the closure error to all parts of the traverse, thereby attempting to approximate the true orientation of the lines and angles in the traverse.

Whereas greater accuracy in this correction may be achieved by more involved methods, such as least-squares techniques, these are likely to be beyond the scope of the FE/EIT exam. Some less computationally demanding techniques using coordinate adjustment are described next.

Compass Rule

This method for traverse closure is appropriate when accuracy of angular measurements is about the same as accuracy of distance measurements. The closure error has two components in Cartesian coordinates—the northing error (δy) and the easting error (δx). In this method, the coordinate error is distributed in proportion to the length of traverse lines. The assumption is that the greatest error will come from the longest shots.

Thus, every line in the traverse has its northing and easting adjusted according to

$$\text{Northing adjustment} = \frac{L_i}{\sum L_i} \times \delta y$$

where L_i is the length of the line being adjusted and δy is the northing closure error.

Easting and elevation values may also be adjusted using the same concept.

Transit Rule

In this method, the coordinate error is distributed in proportion to the amount that various coordinates change between points. Thus, latitudes and departures of lines are adjusted, rather than their coordinates. This method is appropriate when accuracy of angular measurements is much better than accuracy of distance measurements:

$$\text{Latitude adjustment} = \frac{|LAT_i|}{\sum |LAT_i|} \times \delta y$$

$$\text{Departure adjustment} = \frac{|DEP_i|}{\sum |DEP_i|} \times \delta x$$

where LAT_i is the latitude (y-projection) of the line being adjusted and δy is the northing closure error of the traverse, and DEP_i is the departure (x-projection) of the line being adjusted and δx is the easting closure error of the traverse.

Application of the Transit Rule

Let us consider the steps in applying the transit rule, using the data shown in Table 8.2 for a traverse, ABCD. The second and third columns of the table show lengths and azimuthal angles for lines in the traverse. Latitude and departure calculated values are given in the fourth and fifth columns.

First, we need to calculate the closure error for latitude and departure. This is found by summing the latitude and departure values of all the lines. In a perfect closed traverse, these values would sum to zero, but as you can see, the latitude and departure closure errors are 0.054 feet and 0.492 feet, respectively.

The negative of the closure error is the **closure correction**, and it must be distributed in proportion to each segment. So, the next step is to determine the appropriate proportions. The proportion to use is the ratio of the absolute latitude (or departure) of each segment to the sum of these values for all lines in the traverse. The results of these calculations are shown in the sixth and seventh columns of Table 8.2.

Next, the closure correction is distributed proportionally for each line. These calculations are shown in the final two columns of Table 8.2.

Finally, the corrections are added to the original latitudes and departures to obtain corrected values, as shown in Table 8.3.

Solving the same problem using the compass rule, calculations are shown in Table 8.4. The only difference between the two methods is the way in which

Table 8.2 Data for transit rule example

Line	L Length (ft)	u Azimuth Angle	Latitude (ft)	Departure (ft)	$\frac{\|LAT_i\|}{\sum \|LAT_i\|}$	$\frac{\|DEP_i\|}{\sum \|DEP_i\|}$	ΔLAT	ΔDEP
AB	890.32	51° 50′ 40″	+550.039	+700.091	0.1774	0.2916	−0.010	−0.144
BC	1392.85	158° 56′ 40″	−1299.853	+500.413	0.4193	0.2085	−0.023	−0.103
CD	1079.35	256° 36′ 00″	−250.137	−1049.966	0.0807	0.4374	−0.004	−0.215
DA	1011.20	351° 28′ 00″	+1000.006	−150.047	0.3226	0.0625	−0.017	−0.031
			+0.054	**+0.492**			−0.054	−0.492

Table 8.3 Adjusted latitude and departure values after applying the transit rule

Line	Latitude (ft)	Departure (ft)
AB	+550.029	+699.947
BC	−1299.876	+500.311
CD	−250.141	−1050.181
DA	+999.988	−150.077
	Σ = 0.000	Σ = 0.000

Table 8.4 Data for compass rule example

Line	L Length (ft)	u Azimuth Angle	Latitude (ft)	Departure (ft)	$\dfrac{L_i}{\sum L_i}$	ΔLAT	ΔDEP
AB	890.32	51° 50′ 40″	+550.039	+700.091	0.2036	−0.011	−0.100
BC	1392.85	158° 56′ 40″	−1299.853	+500.413	0.3184	−0.017	−0.157
CD	1079.35	256° 36′ 00″	−250.137	−1049.966	0.2468	−0.013	−0.121
DA	1011.20	351° 28′ 00″	+1000.006	−150.047	0.2312	−0.013	−0.114
	Σ = 4373.72		**−0.054**	**−0.492**		**−0.054**	**−0.492**

the ratio is calculated for distributing the correction. For the compass rule, this ratio is calculated from the lengths of the lines AB, BC, CD, and so on, rather than in terms of the latitudes and departures as in the transit rule. Table 8.5 shows the corrected values.

Table 8.5

Line	Latitude (ft)	Departure (ft)
AB	+550.027	+699.991
BC	−1299.870	+500.257
CD	−250.151	−1050.087
DA	+999.993	−150.160
	Σ = 0.000	Σ = 0.000

AREA OF A TRAVERSE

In this section, we will review two methods of calculating the area of a traverse: by coordinates and by double meridian distance.

By Coordinates

To calculate the area of a closed traverse ABCD by coordinates, the formula is

$$A = \left| \sum \frac{1}{2} y_i (x_{i-1} - x_{i+1}) \right|$$

Coordinates x_i and y_i of points A, B, C, and D are often given as eastings and northings, respectively.

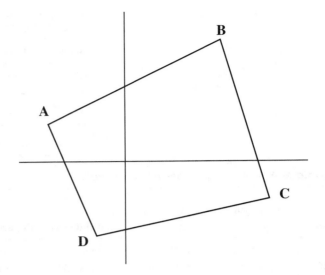

For example, for a closed traverse ABCD, columns 2 and 3 of Table 8.6 specify the coordinates of vertices A, B, C, and D.

Therefore, A = 2,272,500 ÷ 2 = 1,136,250 sq ft = 26.085 acres

Table 8.6 Data for area of traverse by coordinates

Station	Y Northing (ft)	X Easting (ft)	$X_{i-1} - X_{i+1}$	$Y_i (X_{i-1} - X_{i+1})$
A	+500	−250	−100 − 450	−275,000
B	+1050	+450	−250 − 950	−1,260,000
C	−250	+950	−450 − (−100)	−137,500
D	−500	−100	−950 − (−250)	−600,000
				−2,272,500

By Double Meridian Distance

To calculate the area of a closed traverse ABCD by double meridian distance, the formula is

$$A = \left| \sum \frac{1}{2} LAT_i DMD_i \right|$$

where the double meridian distance (DMD) is given by

$$DMD_i = DMD_{i-1} + D_{i-1} + D_i$$

For example, for a closed traverse ABCD, Table 8.7 specifies the latitudes and departures of lines AB, BC, CD, and DA.

Therefore, A = 2,272,500 ÷ 2 = 1,136,250 sq ft = 26.085 acres

Table 8.7 Data for area of traverse by double meridian distance

Segment	ΔY Latitute (ft)	ΔX Departure (ft)	$DMD_i = DMD_{i-1} + D_{i-1} + D_i$	$LAT_i \times DMD_i$
AB	+550	+700	0 + (−150) + 700 = 550	550 × 550 = 302,500
BC	−1300	+500	550 + 700 + 500 = 1750	−1300 × 1750 = −2,275,000
CD	−250	−1050	1750 + 500 − 1050 = 1200	−250 × 1200 = −300,000
DA	+1000	−150	1200 − 1050 − 150 = 0	1000 × 0 = 0
				−2,272,500

Note that for the very first line in the traverse, the preceding DMD value has to be assumed. The assumed value is arbitrary and immaterial. The final DMD value, once calculated, must equal the initially assumed value. Obviously, an assumed value of zero reduces the number of computed terms by one, as the last of the LAT × DMD products becomes zero.

In the preceding example, the first line in the traverse is AB. The preceding line is DA. However, for the purpose of calculating the DMD for AB, the DMD for DA has been assumed to be zero. Upon calculating the DMD for DA, it must therefore also be zero.

AREA UNDER AN IRREGULAR CURVE

When calculating area under a curve (with known vertical ordinates), or volume of earthwork (with known end areas), several numerical schemes are available. Of these, the trapezoidal rule and Simpson's rule are commonly used. A requirement of both formulas is that the stations (marked as $i = 0, 1, 2$, etc., in Figure 8.6) be spaced equal distance apart (shown as Δ in the following equations). Also, for Simpson's rule, the number of intervals (n) must be even.

Note that in Figure 8.6 the first station is marked as $i = 0$ (rather than $i = 1$). This is significant for Simpson's rule, because it involves different weighting factors for odd and even i values (4 and 2, respectively).

Using **linear** approximation between (regularly spaced) nodes, we have the **trapezoidal rule:**

$$A = \frac{\Delta}{2}\left[y_0 + y_n + 2\sum_{i=1}^{n-1} y_i\right]$$

Using quadratic approximation between (regularly spaced) nodes, we have **Simpson's rule** (n must be even):

$$A = \frac{\Delta}{3}\left[y_0 + y_n + 4\sum_{\substack{\text{odd}\\i}} y_i + 2\sum_{\substack{\text{even}\\i}} y_i\right]$$

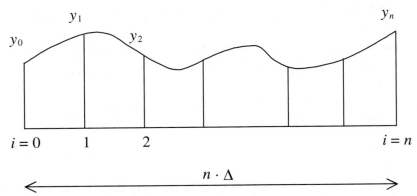

Figure 8.6 Calculating area under an irregular curve

Table 8.8 Sample data for calculating area under a curve

i	x_i (m)	y_i (m)
0	0.0	2.6
1	0.5	3.5
2	1.0	2.4
3	1.5	4.2
4	2.0	4.1
5	2.5	2.0
6	3.0	3.0

For the data shown in Table 8.8, we apply the trapezoidal rule as follows:

$$A = \frac{0.5}{2}[2.6 + 3.0 + 2(3.5 + 2.4 + 4.2 + 4.1 + 2.0)]$$
$$= 9.5 \text{ m}^2$$

For the same data, applying Simpson's rule, we have:

$$A = \frac{0.5}{3}[2.6 + 3.0 + 4(3.5 + 4.2 + 2.0) + 2(2.4 + 4.1)]$$
$$= 9.57 \text{ m}^2$$

The same algorithms may also be used to compute earthwork volumes, given end areas at regularly spaced stations (Figure 8.7). In such a case, the end areas serve as ordinates (y), whereas the station coordinates serve as x values. Using the average end area method, the volume between two stations may be expressed as:

$$V = L\left(\frac{A_1 + A_2}{2}\right)$$

When calculating volumes at a location where a fill transitions to a cut (or vice versa), we might encounter regions where one of the end areas is negligible (Figure 8.8). For better accuracy, such volumes should be calculated using the pyramid formula, rather than the average end area method. The formula is as follows:

$$V = L\left(\frac{A}{3}\right)$$

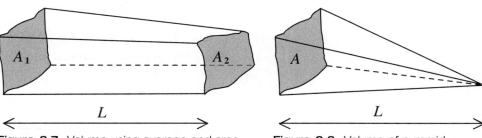

Figure 8.7 Volume using average end area **Figure 8.8** Volume of pyramid

PROBLEMS

8.1 The back-tangent to a horizontal circular curve has a bearing of N 34° 44′ 35″W. The deflection angle between tangents is 67° to the right. If the PC is located at coordinates 4453.51 m N, 643.29 m W, and the tangent length is 850.32 m, what are the coordinates of the PT?
  a. 5657.19 N, 888.18 W
  b. 5850.95 N, 643.29 W
  c. 5871.32 N, 674.06 W
  d. 4184.70 N, 147.33 E

8.2 The following data were obtained as foresight and backsight angles at a station in a traverse. What is the horizontal angle at this station?

		Plate Reading
Backsight	Direct reading	00° 00′ 00″
	Reverse reading	180° 00′ 05″
Foresight	Direct reading	51° 40′ 22″
	Reverse reading	231° 40′ 16″

  a. 51° 40′ 22″
  b. 51° 40′ 16.5″
  c. 51° 40′ 11″
  d. 51° 40′ 15″

8.3 The table below describes a traverse ABCDE. Determine the interior angle at C.

Line	Azimuth Angle	Bearing	Length (m)	Deflection
AB	132	–	237.12	–
BC	–	–	156.18	47° 30′ left
CD	158° 02′ 30″	–	349.65	–
DE	–	–	165.76	105° 16′ 45″ left

  a. 73° 32′ 30″
  b. 106° 27′ 30″
  c. 74° 43′ 15″
  d. 84° 30′ 00″

8.4 For the traverse shown below, calculate the coordinates of station D if the coordinates of A are (562.34 m N, 760.27 m W).

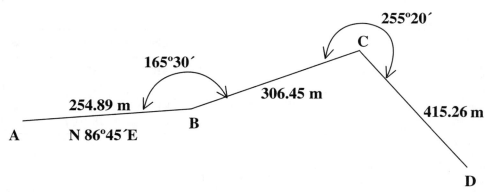

Exhibit 8.4

a. 1331.29 N, 1002.94 W
b. 319.67 N, 1002.94 E
c. 1331.29 N, 8.68 W
d. 319.67 N, 8.68 E

8.5 Two tangents intersecting at an angle of 35° 45′ are to be joined by a horizontal circular curve. If the degree of curve (based on a 100-ft arc) is 9° 30′, compute the tangent distance and the radius of the curve.
a. 185.12 ft
b. 194.51 ft
c. 574.00 ft
d. 94.85 ft

8.6 The table below shows differential leveling data using a transit level. The starting station is of known elevation. Find the elevation of station D.

Station	B.S (m)	F.S. (m)	Elevation (m)	Notes
A	3.95	–	500.00	Benchmark
B	2.47	6.34		
C	3.81	5.51		
D	–	6.78		

a. 508.40 m
b. 489.77 m
c. 510.23 m
d. 491.60 m

8.7 The table below shows length and azimuthal angles for lines in a closed traverse, ABCD. What is the correction to the departure of CD, using the transit rule?

Line	Length (ft)	Azimuth Angle
AB	850.00	80° 30′
BC	1250.00	136° 15′
CD	1000.00	220° 30′
DA	1850.00	325° 20′

a. +0.192 ft
b. − 0.192 ft
c. − 0.343 ft
d. +0.343 ft

8.8 Assuming the earth to be a perfect sphere of radius 6370 km, what is the curved distance (measured on the surface) between two points 1° of longitude apart on the 32nd parallel (32° N or 32° S latitude)?
a. 94,284 m
b. 42,371 m
c. 84,732 m
d. 47,142 m

SOLUTIONS

8.1 c. Bearing angle for forward tangent = 67° − 34° 44′ 35″ = N32° 15′ 25″E (32.257 deg)

Coordinates of PI:

Northing = 4453.51 + (850.32)cos(34.743°) = + 5152.23

Easting = − 643.29 − (850.32)sin(34.743°) = − 1127.88

Coordinates of PI: 5152.23 N, 1127.88 W

Coordinates of PT:

Northing = 5152.23 + (850.32)cos(32.357°) = + 5871.32

Easting = − 1127.88 + (850.32)sin(32.357°) = − 674.06

Coordinates of PT: 5871.032 N, 674.06 W

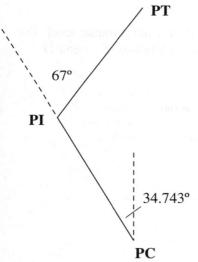

Exhibit 8.1

8.2 b. Backsight readings differ by 180° 00′ 05″

Error 00° 00′ 05″

Adjusted backsight readings:

Direct: 00° 00′ 02.5″

Reverse: 180° 00′ 02.5″

Foresight readings differ by 179° 59′ 54″

Error 00° 00′ 06″

Adjusted foresight readings:

Direct: 51° 40′ 19″

Reverse: 231° 40′ 19″

Corrected Angle 51° 40′ 19″ −00° 00′ 02.5″ = 51° 40′ 16.5″

8.3 b. The azimuthal angles for all lines may be calculated starting with AB:

Az(AB) = 132° (given)

Az(BC) = 132° − 47° 30′ = 84° 30′ (Bearing BC = N 84° 30′ 00″ E)

Az(CD) = 158° 02′ 30″ (given)   (Bearing BC = S 21° 57′ 30″ E)

Az(DE) = 158° 02′ 30″ − 105° 16′ 45″ = 52° 45′ 45″

Interior angle at C = Azimuth(BC) − Azimuth(CD) + 180° = 84° 30′ − 158° 02′ 30″ + 180° = 106° 27′ 30″

8.4 d.

Line	Length (m)	Azimuth (degrees)	Lat = $L\cos(Az)$ (m)	Dep = $L\sin(Az)$ (m)
AB	254.89	86.75	+14.45	+254.48
BC	306.45	72.25	+93.43	+291.86
CD	415.26	147.58	−350.55	+222.61

Starting with coordinates of A (+ 562.34, − 760.27), we obtain the coordinates of D as follows:

$$562.34 + 14.45 + 93.43 − 350.55 = + 319.67$$
$$− 760.27 + 254.48 + 291.86 + 222.61 = + 8.68$$

Coordinates of D are (319.67 N, 8.68 E).

8.5 b. $I = 35°45' = 35.75°$
$D = 9°30' = 9.5°$
$R = \dfrac{5729.578}{9.5} = 603.11 \text{ ft}$
$T = R \tan \dfrac{I}{2} = 603.11 \times \tan 17.875° = 194.51 \text{ ft}$

8.6 d. Sum of all backsight values = 10.23

Sum of all foresight values = 18.63

Elevation of D = Elevation of A + Sum of backsight − Sum of foresight
= 491.60 m

8.7 b. Sum of departures = $\Sigma(L \sin \theta)$ = 850sin80.5° + 1250sin136.25° + 1000sin220.5° + 1850sin325.33° = 838.343 + 864.391 − 649.448 − 1052.282 = + 1.004

Departure correction for CD = $\dfrac{649.448}{838.343 + 864.391 + 649.448 + 1052.282}$ × −1.004 = −0.192 ft

8.8 a. If the equatorial radius is R, the radius at a latitude θ is $R\cos\theta$.

The circumference of this small circle is $2\pi R\cos\theta$, which is divided into 360 degrees of longitude.

Thus, at this latitude, each degree of longitude is equivalent to a distance of $2\pi R\cos\theta \div 360$.

The answer is 94.284 km.

CHAPTER 9

Construction Management

Thomas Nelson

OUTLINE

PROCUREMENT METHODS 122
Design/Bid/Build ■ Design/Build ■ Construction Manager

CONTRACT TYPES 124
Lump Sum ■ Unit Price ■ Cost Plus

CONTRACTS AND CONTRACT LAW 124
Bidding ■ Bonding

CONSTRUCTION ESTIMATING 125
Soil Volume Changes ■ Spoil Banks/Piles ■ Pit Excavation ■ Linear Cut/Fill

PRODUCTIVITY 129

PROJECT SCHEDULING 129
Gantt/Bar Charts ■ Critical Path Method (CPM) ■ Program Evaluation and Review Technique (PERT)

PROBLEMS 136

SOLUTIONS 137

Approximately 10% of the civil FE/EIT exam concerns construction management topics; candidates can expect to see six construction management questions. According to NCEES, the following specific topics may be covered:

- Procurement methods
- Allocation of resources
- Contracts and contract law
- Project scheduling
- Engineering economics

- Project management
- Construction estimating

This chapter covers key terms, concepts, and techniques of construction management that you may encounter on the exam. Coverage of engineering economics, which is also tested in the morning portion of the FE/EIT exam, can be found in Chapter 15 of *Fundamentals of Engineering: FE/EIT Exam Preparation*, 18th edition, published by Kaplan AEC Education.

PROCUREMENT METHODS

Three important procurement methods in construction management are as follows:

1. Design/bid/build
2. Design/build
3. Construction manager

Design/Bid/Build

Design/bid/build is the conventional procurement method that solicits bids from contractors for construction, following completion of the design by an architecture/engineering (AE) firm. (See Figure 9.1.) It is a sequential process, with contractors bidding on a complete set of plans. Most often the lowest bid is sought. The client establishes separate contracts with the AE and the contractor for their work. There is no contractual relationship between the AE and contractor, although the client may hire the AE to perform inspection functions.

Design/Build

The design/build method selects one firm to both design and build the project. (See Figure 9.2.) It is usually advantageous for the client to deal with one firm for

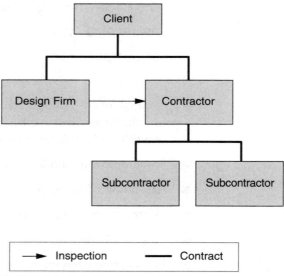

Figure 9.1 Relationships in the design/bid/build procurement method

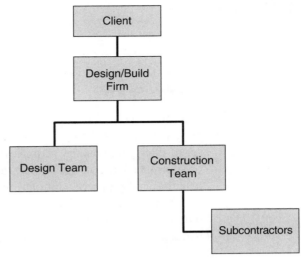

Figure 9.2 Relationships in the design/build procurement method

both design and construction, often termed "turn-key." Disputes between designer and contractor are handled within the one firm. Coordination is also better. One major advantage is that design and construction can be done concurrently. This is often called "fast track" or "phased construction," meaning that construction on some aspects of the project can begin while other aspects are still in design. Design/build is often used on large, complex projects with tight time schedules or for those that are not completely defined initially. The disadvantage to the client is that he or she does not know the total cost until the project is complete.

Construction Manager

The construction manager method uses a construction manager (CM), hired by the client to manage the work of the AE and the contractor, who each have contracts with the client. (See Figure 9.3.) This usually results in time and cost savings due to the close supervision of the design and construction efforts.

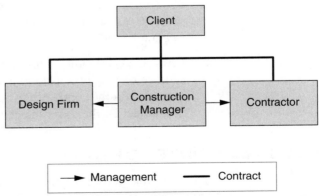

Figure 9.3 Relationships in the construction manager procurement method

CONTRACT TYPES

In this section, we will review three main contract types for construction management: lump sum, unit price, and cost plus.

Lump Sum

A lump sum contract stipulates a total price for all work to be performed, including all material, labor, equipment, overhead, and profit. This contract requires a detailed set of plans and quantity estimates. Any difference in costs requires a change order, or a formal change in the contract, agreed on by both parties. Lump sum contracts are usually used with conventional procurement.

A lump sum contract is advantageous to the client, who knows exactly how much the project will cost. In addition to requiring a detailed set of plans, a disadvantage is that changes are more difficult to make once construction has started.

Unit Price

A unit price contract is used when quantities may not be known, often with projects that include large amounts of excavation. The contractor bids a price for each work item by the cost per unit (for example, cost per cubic yard of excavation). The client then agrees to pay the contractor for each unit of work performed. A low bid can be used, based on the contractor's bidding on an engineer's estimated quantities. Actual quantities determine the total price paid to the contractor. All unit price bids should include any overhead and profit.

A unit price contract is an excellent one to use when exact quantities are unknown, but it requires agreement between contractor and client as to actual quantities of work performed. Adjustments are usually made to the unit price whenever large differences in quantities are encountered.

Cost Plus

A *cost plus* contract is used when the total scope of work is not known. The client agrees to pay the contractor for the actual work performed and for overhead. The contractor's profit is the *plus* portion of the contract. This contract type is useful for emergency construction or for projects that must be started before design is complete. The *plus* portion of the terms can have several variations, including the following:

- *Cost plus fixed fee*. This contract pays the contractor for costs and overhead plus a fixed amount for markup/profit, which gives the contractor incentive to complete the project as quickly and economically as possible.

- *Cost plus percentage*. This contract pays the contractor for costs plus a percentage of the costs as markup/profit. Under this type contract, there is no incentive for the contractor to be efficient. On the contrary, the more expensive the project is, the more the contractor makes.

CONTRACTS AND CONTRACT LAW

The morning portion of the FE exam may contain some general questions on agreements and contracts. Here, we will focus on legal issues specific to construction management.

Bidding

The bid process begins with a Notice to Bidders, indicating the scope and location of the project, client, availability of plans and specifications to be used, date and location of submission, and bond requirements. Bidders may be required to attend a prebid meeting/site visit to ask questions, clarify scope, and see the actual project site.

Interested contractors will estimate material and labor costs, time and equipment requirements, and overhead and profit desired. The completed bid must be submitted at the proper location no later than the time required. Late submissions will be rejected.

Bonding

On all public construction projects, and most private projects, three types of bonds are obtained by contractors prior to submitting a bid and are obtained from a surety company:

1. *Bid bond.* Typically 5% to 20% of estimated project cost. The bid bond guarantees that the bidder will enter into a contract with the client if the bidder is the low bidder. If the low bidder does not sign a contract, the amount of the bid bond is used by the client to offset either the cost of the next lowest bid or the cost of rebidding the project.

2. *Performance bond.* Typically 100% of project cost. This bond guarantees that the contractor will perform the specified work in accordance with the contract. If the contractor defaults on the contract, the bond is the upper amount that the surety company will incur to arrange completion of the project.

3. *Payment (labor and material payment) bond:* Typically 100% of project cost. The payment bond guarantees that the contractor will pay for all materials and labor used on the project, protecting the client from liens against the project by third parties.

CONSTRUCTION ESTIMATING

In preparing a bid, one of the most commonly estimated items is earthwork excavation. The following sections present equations used in estimating earthwork quantities.

Soil Volume Changes

In earthwork operations, material will change in volume depending on its position in the construction operation. A given weight of soil will occupy different volumes depending on whether it is in its natural (bank) condition, loose, or compacted (Table 9.1). Generally, soils will swell 30% to 40% between natural and loose

Table 9.1 Soil volume differences

Bank yards	Density = 2000 lb/yd^3 = 2000 lb/yd^3
Loose yards (30% swell)	Density = 2000 lb/1.3yd^3 = 1538 lb/yd^3
Compacted yards (0.75 shrinkage factor)	Density = 2000 lb/0.75 yd^3 = 2667 lb/yd^3

Table 9.2 Soil unit weights and change factors

	Unit Weight (lb/cu yd)			Swell (%)	Shrinkage (%)	Swell Factor	Shrinkage Factor
	Bank	Loose	Compacted				
Clay	3000	2310	3750	30	20	0.77	0.80
Earth	3100	2480	3450	25	10	0.80	0.90
Rock, loose	4600	3060	3550	50	−30*	0.67	1.30*
Sand/gravel	3200	2860	3650	12	12	0.89	0.88

* Compacted rock is not as dense as bank rock.

conditions and will shrink 10% to 20% from natural to compacted condition. The quantities of earth in the process of excavating from in-place to loose to compacted condition are measured in bank cubic yards (BCY), loose cubic yards (LCY), and compacted cubic yards (CCY).

Swell factors and shrinkage factors are used to determine the volume changes. Swell factor measures the increase in volume from natural (bank) to loose condition (e.g., from natural to dump truck). Swell factor is often termed the load factor and is defined as:

Swell factor = Loose unit weight ÷ Bank unit weight
Swell factor = 1 ÷ (1 +swell)

Loose volume is multiplied by the swell factor to obtain the bank volume.

The shrinkage factor measures the decrease in volume from bank to compacted condition. It is defined as:

Shrinkage factor = Bank unit weight ÷ Compacted unit weight
Shrinkage factor = 1 − Shrinkage

Bank volume is multiplied by the shrinkage factor to obtain the compacted volume.

Typical soil unit weights and change factors are summarized in Table 9.2.

Spoil Banks/Piles

Excavated (loose) material needs to be stored in either banks (triangular cross section) or piles (conical shape). The volume of the spoil bank/pile is measured in loose cubic yards (LCY). Dimensions can be calculated if the angle of repose (angle between the slope and horizontal plane) of the soils is known. Table 9.3 lists some typical angles of repose.

Table 9.3 Angles of repose for common excavation materials

Material	Angle of Repose (R) in Degrees
Clay	35
Common earth, dry	32
Common earth, moist	37
Gravel	35
Sand, dry	25
Sand, wet	37

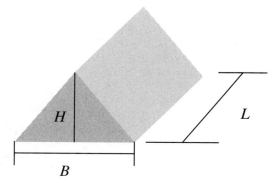

Figure 9.4 A triangular spoil bank

For a triangular spoil bank, as shown in Figure 9.4, relevant equations are as follows:

$$\text{Volume} = \text{Section area} \times \text{Length}$$
$$\text{Volume} = (\tfrac{1}{2}BH)\,L$$
$$B = [(4V \div (L \times \tan R)]^{1/2}$$
$$H = (B \times \tan R) \div 2$$

Or,

$$\text{Volume} = (B^2\,L\,\tan R)/4$$

where
 B = base width (ft)
 H = pile height (ft)
 L = pile length (ft)
 R = angle of repose (degrees)
 V = volume (cu ft)

For a conical spoil pile, as shown in Figure 9.5, the relevant equations are:

$$\text{Volume} = 1/3 \times \text{Base area} \times \text{Height}$$
$$D = (24V/\pi \tan R)^{1/3}$$
$$H = D/2 \times \tan R$$

where
 D = diameter of pile base (ft)
 H = height of pile (ft)

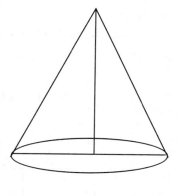 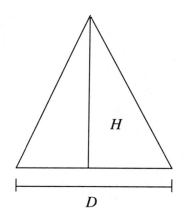

Figure 9.5 Conical spoil pile

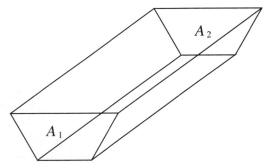

Figure 9.6 Geometry for the average end area method

Pit Excavation

In order to calculate the volume of excavation, the surface area is divided into regular-sized grids. The depth of excavation is determined for each corner of each grid square by subtracting the cut elevation from the surface elevation. The volume of excavation is found by:

Volume (BCY) = Area of surface (sq ft) × Average depth of cut (ft)

Linear Cut/Fill

Road excavation (cut) or material added (fill) are calculated by 100–foot station. The volume of cut or fill is calculated by the average end area or the trapezoidal methods.

Average End Area Method

Figure 9.6 shows the geometry for the average end area method of calculating cut/fill volume.

$$\text{Volume (cu yd)} = \frac{(A_1 + A_2)}{2} (L) \frac{1 \text{ yd}^3}{27 \text{ ft}^3}$$

where
A_1 and A_2 = area in sq ft for respective end areas
L = length between end areas (ft)

Prismoidal Method

Figure 9.7 shows the geometry for the prismoidal method of calculating cut/fill volume.

$$\text{Volume (cu yd)} = \frac{(A_1 + 4A_m + A_2)}{6} (L) \frac{1 \text{ yd}^3}{27 \text{ ft}^3}$$

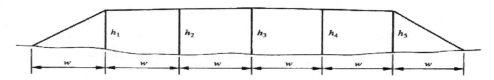

Figure 9.7 Geometry for the prismoidal method

where

A_1 and A_2 = end areas
A_m = area at the midpoint of the length

The end areas can be determined by breaking the cross section into triangles and trapezoids, determining the area of each and summing the area:

$$\text{Area of triangle} = hw/2$$

where
h = height of triangle
w = width of triangle

$$\text{Area of trapezoid} = (h_1 + h_2)/2 \times w$$

where
h_1 and h_2 = lengths of parallel sides
w = distance between parallel sides

The general rule for calculation of multiple trapezoidal areas is:

$$\text{Area} = w(h_0/2 + h_1 + h_2 + \ldots + h_{(n-1)} + h_n/2)$$

In general, the prismoidal method will produce a more accurate volume than the average end area method. It is common that excavation be expressed in BCY, hauling in LCY, and final volume in CCY. It is important to work in one measure and then convert to the required condition.

PRODUCTIVITY

Equipment productivity can be calculated by:

$$\text{LCY/hr} = \text{Cycles per hour} \times \text{Bucket payload (LCY) per cycle}$$

The payload is calculated by multiplying the bucket size by a fill factor, which ranges from 0.40 to 1.10, depending on the type of material being loaded.

PROJECT SCHEDULING

Several scheduling techniques are commonly used on construction projects. In this review, we will focus on Gantt/bar charts, CPM, and PERT.

Gantt/Bar Charts

Gantt/bar charts are the simplest of scheduling techniques. They indicate each task or work by start and end date. The chart may indicate milestones and percent completion of each task. It does not indicate relationships between activities. Figure 9.8 shows a sample bar chart.

Critical Path Method (CPM)

CPM diagrams may be either activity-on-arrow (AOA) or activity-on-node (AON). Both methods indicate the logical relationship between activities and can be used to indicate the shortest completion time and the activities that must be completed on time to ensure the timely completion of the project, the critical path. Calculations

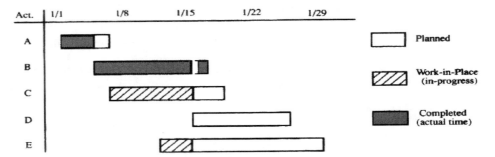

Figure 9.8 A generic Gantt/bar chart

will also determine how much time activities may be delayed without delaying either the entire project (total float) or the succeeding activities (free float).

Activity-on-Arrow (AOA)

In AOA notation, each activity is represented by an arrow. Activities may be labeled with a name or with the numbers of the nodes at the start and end of the activity. This is known as *ij* notation. Each activity has a duration measured in days and begins and ends at a node.

The notation shown in Figure 9.9 is as follows:

- ES = early start, the earliest the activity can start (all immediate preceding activities completed)

- EF = early finish, the earliest the activity can finish = ES + duration

- LS = late start, the latest the activity can start without delaying project completion

- LF = late finish, the latest the activity can finish without delaying project completion

- TF = total float, the amount the activity can be delayed without delaying completion of project

- FF = free float, the amount the activity can be delayed without delaying succeeding activities

No activity leaving a node can be started until all activities entering the node have been completed. The CPM diagram in Figure 9.10 shows a project with five activities. The duration for each activity is shown below the activity letter. Activity d1 is a *dummy* activity, which has a duration of zero days but shows a logical dependence of activity E not being able to start until activity B is completed (as well as activity C).

In order to determine the critical path, a forward pass through the CPM diagram is performed to calculate the ES and EF:

$$ES = \text{latest EF of all immediate preceding activities}$$
$$EF = ES + \text{Duration}$$

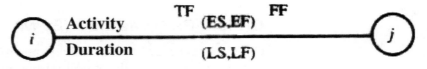

Figure 9.9 *ij* notation from a CPM diagram

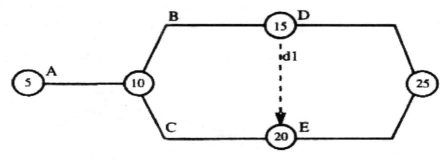

Figure 9.10 CPM diagram—activity-on-arrow method

This is followed by a backward pass to calculate the LS and LF:

LF = earliest LS of all succeeding activities
LS = LF − Duration

Total float is then calculated as TF = LS − ES. Finally, free float is calculated as the difference between the ES of each succeeding activity minus the EF of the activity (smallest value is selected).

Example 9.1

Using the CPM diagram shown in Figure 9.10 and the following activity durations, determine the project duration, critical path, and float.

Activity	Duration	Activity	Duration
A	5	D	2
B	4	E	4
C	3		

Solution

Calculations for activity E are shown as an example.

ES = 9 (Activity C finishes at end of day 8, but activity B, with dummy, ends at end of day 9)

EF = ES + Duration = 9 + 4 = 13

LF = 13 (the ES of succeeding activity, which is project end)

LS = LF − Duration = 13 − 4 = 9

TF = LS − ES = 9 − 9 = 0

FF = 13 − 13 = 0 (ES of project end is 13; EF of E is 13)

Using the CPM diagram in Figure 9.10, a table is set up to record all values:

Activity	Duration	ES	EF	LS	LF	TF	FF	Critical?
A	5	0	5	0	5	0	0	*
B	4	5	9	5	9	0	0	*
C	3	5	8	6	9	1	1	
D	2	9	11	11	13	2	2	
E	4	9	13	9	13	0	0	*

* Indicates activities on the critical path

Any activity with a total float of zero is on the critical path. Activities with a total not equal to zero can be delayed that number of days without delaying the project. Activities with free float not equal to zero can be delayed that number of days until the next activity is delayed. Free float does not have to equal total float.

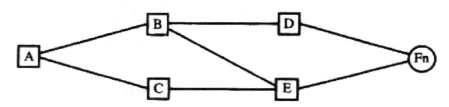

Figure 9.11 CPM diagram—activity on node methodŽ

Activity-on-Node (AON)

An alternate method of drawing the CPM diagram is to use activity on node. In this method, activities are represented by the node, not by the arrow. Nodes may be a circle (bubble diagram) or a rectangle (often termed precedence). AON is popular because most scheduling software uses this method. The AON diagram in Figure 9.11 shows the same project as the AOA diagram in Figure 9.10.

Calculations are performed in the same manner as with the AOA diagram. Instead of placing time calculation information above and below the arrow, information is usually placed inside the node, as shown in Figure 9.12.

One advantage of precedence diagrams is that the location of the tail and head of the line connecting two activity nodes can indicate more detailed dependencies between the two activities. This is shown in Table 9.4.

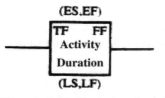

Figure 9.12 Typical node labeling

Program Evaluation and Review Technique (PERT)

PERT uses the CPM logic and statistics to determine the most likely time of completion for a project. Each activity has an optimistic completion time, pessimistic completion time, and a most likely completion time. For each activity, an expected completion time (t_e) is calculated, along with a standard deviation (σ) and a variance (v). Using normal distribution tables, a probable completion date is calculated as follows:

$$t_e = (a + 4m + b)/6$$
$$\sigma_{te} = (b - a)/6$$
$$v = \sigma_{te}^2$$

where
 a = optimistic activity duration
 m = most likely duration
 b = pessimistic activity duration

The variance (V) for the critical path is the sum of the variances for the activities on the critical path, and $\sigma_{TE} = V^{1/2}$.

The critical path is assumed to have a normal distribution. Therefore, the probability of completion of a project can be determined using a Z table as follows (see Table 9.5):

$$Z = (T_s - T_E)/\sigma_{TE}$$

Table 9.4 Node configurations show activity dependencies

CPM Node Diagram	Relationship	Description
A → B (vertical)	Start-to-start	The start of B depends on the start of A.
A → B (vertical, finish)	Finish-to-finish	The finish of B depends on the finish of A.
A → B (horizontal)	Finish-to-start	The start of B depends on the finish of A.
A → B (combination)	Combination	The start and finish of B depend on the start and finish of A.

Wb

where
    Z = number of standard deviations from the mean
    T_E = mean of critical path
    σ_{TE} = standard deviation of critical path = $V^{1/2}$
    T_s = date of interest

The CPM diagram shown in Figure 9.13 shows three durations for each activity. The first is the optimistic time (a), the second is the most likely (m), and the third is the pessimistic time (b).

The expected duration (t_e) is found as:

$$t_e = (2 + 4 \times 5 + 14)/6 = 36/6 = 6$$

The standard deviation (σ_{te}) is found as:

$$\sigma_{te} = (14 - 2)/6 = 2$$

Table 9.5 A Z table

Z	P, probability of completing by T_s	Z	P, probability of completing by T_s
−3.0	0	+0.1	.54
−2.5	.01	+0.2	.58
−2.0	.02	+0.3	.62
−1.5	.07	+0.4	.66
−1.4	.08	+0.5	.69
−1.3	.10	+0.6	.73
−1.2	.12	+0.7	.76
−1.1	.14	+0.8	.79
−1.0	.16	+0.9	.82
−0.9	.18	+1.0	.84
−0.8	.21	+1.1	.86
−0.7	.24	+1.2	.88
−0.6	.27	+1.3	.90
−0.5	.31	+1.4	.92
−0.4	.34	+1.5	.93
−0.3	.38	+2.0	.98
−0.2	.42	+2.5	.99
−0.1	.46	+3.0	1.00
0	.50		

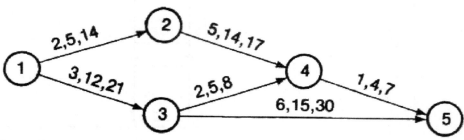

Figure 9.13 An example CPM

The variance is:

$$v = \sigma_e^2 = 2^2 = 4$$

Performing the same calculations for all the activities yields the following:

Activity	t_e	s_{te}	n
1,2	6	2	4
1,3	12	3	9
2,4	13	2	4
3,4	5	1	1
4,5	4	1	1
3,5	16	4	16

Using the expected durations for each activity (t_e) and the CPM, the critical path is 1 – 3, 3 – 5, with a duration of 12 + 16 = 28 days. The variance (V) = 9 + 16 = 25.

$$\sigma_{TE} = 25^{1/2} = 5.$$

The probability of completing this project by the end of day 23 is found by:

$$Z = (23 - 28)/5 = -1.0$$

The Z table shows a probability of 0.16 for $Z = -1.0$. Therefore, there is a 16% probability of completing the project by the end of day 23.

PROBLEMS

9.1. A dump truck can haul 15 loose cubic yards of common earth. If the material has a swell factor of 0.80, how many bank cubic yards must be excavated to fill the dump truck?
a. 15 BCY
b. 12 BCY
c. 18.75 BCY
d. 20 BCY

9.2. If the shrinkage of the material in question 9.1 is 10%, how much volume will 100 BCY of this material fill when compacted?
a. 90 CCY
b. 100 CCY
c. 111 CCY
d. 99 CCY

9.3. What is the volume of excavation in a gravel pit, if the depths of cut at the corners are 6.0′, 5.8′, 7.6′, and 8.2′ ? The area is 20 ft by 20 ft.
a. 102.2 BCY
b. 306.6 BCY
c. 102.2 LCY
d. 306.6 LCY

9.4. An excavator will be used to load trucks. The contractor has the choice of four excavators, each having a different cycle time and bucket size. Which excavator is the most productive?

	Cycle Time (min)	Bucket Size (LCY)
Excavator A	0.32	1.00
Excavator B	0.25	0.75
Excavator C	0.50	1.50
Excavator D	0.60	2.00

a. Excavator A
b. Excavator B
c. Excavator C
d. Excavator D

9.5. Using PERT, if a project has an expected duration of 50 days, with a project variance of 16 days, what is the probability of the project being completed by the end of day 48?
a. 31%
b. 69%
c. 50%
d. 55%

9.6. Using the information in question 9.5, on which day will the project be completed with 90% probability?
a. 45
b. 46
c. 55
d. 56

Use the CPM diagram in Exhibit 9.6 to answer questions 9.7, 9.8, and 9.9.

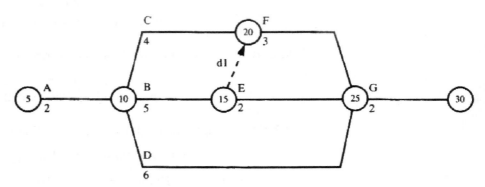

Exhibit 9.6 CPM diagram

9.7. What is the duration of the project?
a. 11 days
b. 12 days
c. 13 days
d. 10 days

9.8. Which activities are on the critical path?
a. A-B-E-G
b. A-D-G
c. A-C-F-G
d. A-B-F-G

9.9. How long can activity D be delayed without delaying the project completion?
a. 1 day
b. 2 days
c. 3 days
d. 0 days

SOLUTIONS

9.1. b. BCY = LCY × Swell factor = 15 × 0.8 = 12 BCY

9.2. a. Shrinkage factor = 1 − Shrinkage = 1 − 0.1 = 0.90
CCY = BCY × Shrinkage factor = 100 × 0.90 = 90 CCY

9.3. a. Average depth of cut = (6.0 + 5.8 + 7.6 + 8.2)/4 = 6.9 ft
Area = 20′ × 20′ = 400 sq ft
Volume = (400 × 6.9)/27 = 102.2 BCY (excavation is measured in BCY)

9.4. d. Excavator A = (60/0.32) × 1 CY = 187.5 LCY

Excavator B = (60/0.25) × 0.75 CY = 180 LCY

Excavator C = (60/0.50) × 1.5 CY = 180 LCY

Excavator D = (60/0.60) × 2.0 CY = 200 LCY

9.5. a. $\sigma_{TE} = V^{1/2} = 16^{1/2} = 4$

$$Z = (48 - 50)/4 = -0.50$$
From Z table, $p = 0.31 = 31\%$

9.6. d. From Z table, working backwards, $p = 0.90$. This gives $Z = 1.3$.

$$Z = 1.3 = (x - 50)/4$$
$$x = 4(1.3) + 50 = 55.2 \text{ during 56th day}$$

9.7–9.9. Sample calculations for activity F are shown.

ES = 7 (Activity C finishes at end of day 6, but activity B ends at end of day 7)

EF = ES + Duration = 7 + 3 = 10

LF = 10 (the ES of succeeding activity, G)

LS = LF − duration = 10 − 3 = 7

TF = LS − ES = 7 − 7 = 0

FF = 10 − 10 (ES of G is 10; EF of F is 10)

Using the CPM diagram shown in Exhibit 9.6, a table is set up to record values for all activities:

Activity	Duration	ES	EF	LS	LF	TF	FF	Critical?
A	2	0	2	0	2	0	0	*
B	5	2	7	2	7	0	0	*
C	4	2	6	3	7	1	1	
D	6	2	8	4	10	2	2	
E	2	7	9	8	10	1	1	
F	3	7	10	7	10	0	0	*
G	2	10	12	10	12	0	0	*

9.7. b. See the preceding calculations.

9.8. d. Activities A-B-F-G have 0 total float, so they are on the critical path.

9.9. b. See the preceding table.

APPENDIX

Afternoon Sample Examination

INSTRUCTIONS FOR AFTERNOON SESSION

1. You have four hours to work on the afternoon session. You may use the *Fundamentals of Engineering Supplied-Reference Handbook* as your *only* reference. Do not write in this handbook.

2. Answer every question. There is no penalty for guessing.

3. Work rapidly and use your time effectively. If you do not know the correct answer, skip it and return to it later.

4. Some problems are presented in both metric and English units. Solve either problem.

5. Mark your answer sheet carefully. Fill in the answer space completely. No marks on the workbook will be evaluated. Multiple answers receive no credit. If you make a mistake, erase completely.

Work 60 afternoon problems in four hours.

FUNDAMENTALS OF ENGINEERING EXAM
AFTERNOON SESSION

Ⓐ Ⓑ Ⓒ Ⓓ Fill in the circle that matches your exam booklet

1. Ⓐ Ⓑ Ⓒ Ⓓ
2. Ⓐ Ⓑ Ⓒ Ⓓ
3. Ⓐ Ⓑ Ⓒ Ⓓ
4. Ⓐ Ⓑ Ⓒ Ⓓ
5. Ⓐ Ⓑ Ⓒ Ⓓ
6. Ⓐ Ⓑ Ⓒ Ⓓ
7. Ⓐ Ⓑ Ⓒ Ⓓ
8. Ⓐ Ⓑ Ⓒ Ⓓ
9. Ⓐ Ⓑ Ⓒ Ⓓ
10. Ⓐ Ⓑ Ⓒ Ⓓ
11. Ⓐ Ⓑ Ⓒ Ⓓ
12. Ⓐ Ⓑ Ⓒ Ⓓ
13. Ⓐ Ⓑ Ⓒ Ⓓ
14. Ⓐ Ⓑ Ⓒ Ⓓ
15. Ⓐ Ⓑ Ⓒ Ⓓ
16. Ⓐ Ⓑ Ⓒ Ⓓ
17. Ⓐ Ⓑ Ⓒ Ⓓ
18. Ⓐ Ⓑ Ⓒ Ⓓ
19. Ⓐ Ⓑ Ⓒ Ⓓ
20. Ⓐ Ⓑ Ⓒ Ⓓ
21. Ⓐ Ⓑ Ⓒ Ⓓ
22. Ⓐ Ⓑ Ⓒ Ⓓ
23. Ⓐ Ⓑ Ⓒ Ⓓ
24. Ⓐ Ⓑ Ⓒ Ⓓ
25. Ⓐ Ⓑ Ⓒ Ⓓ
26. Ⓐ Ⓑ Ⓒ Ⓓ
27. Ⓐ Ⓑ Ⓒ Ⓓ
28. Ⓐ Ⓑ Ⓒ Ⓓ
29. Ⓐ Ⓑ Ⓒ Ⓓ
30. Ⓐ Ⓑ Ⓒ Ⓓ
31. Ⓐ Ⓑ Ⓒ Ⓓ
32. Ⓐ Ⓑ Ⓒ Ⓓ
33. Ⓐ Ⓑ Ⓒ Ⓓ
34. Ⓐ Ⓑ Ⓒ Ⓓ
35. Ⓐ Ⓑ Ⓒ Ⓓ
36. Ⓐ Ⓑ Ⓒ Ⓓ
37. Ⓐ Ⓑ Ⓒ Ⓓ
38. Ⓐ Ⓑ Ⓒ Ⓓ
39. Ⓐ Ⓑ Ⓒ Ⓓ
40. Ⓐ Ⓑ Ⓒ Ⓓ
41. Ⓐ Ⓑ Ⓒ Ⓓ
42. Ⓐ Ⓑ Ⓒ Ⓓ
43. Ⓐ Ⓑ Ⓒ Ⓓ
44. Ⓐ Ⓑ Ⓒ Ⓓ
45. Ⓐ Ⓑ Ⓒ Ⓓ
46. Ⓐ Ⓑ Ⓒ Ⓓ
47. Ⓐ Ⓑ Ⓒ Ⓓ
48. Ⓐ Ⓑ Ⓒ Ⓓ
49. Ⓐ Ⓑ Ⓒ Ⓓ
50. Ⓐ Ⓑ Ⓒ Ⓓ
51. Ⓐ Ⓑ Ⓒ Ⓓ
52. Ⓐ Ⓑ Ⓒ Ⓓ
53. Ⓐ Ⓑ Ⓒ Ⓓ
54. Ⓐ Ⓑ Ⓒ Ⓓ
55. Ⓐ Ⓑ Ⓒ Ⓓ
56. Ⓐ Ⓑ Ⓒ Ⓓ
57. Ⓐ Ⓑ Ⓒ Ⓓ
58. Ⓐ Ⓑ Ⓒ Ⓓ
59. Ⓐ Ⓑ Ⓒ Ⓓ
60. Ⓐ Ⓑ Ⓒ Ⓓ

DO NOT WRITE IN BLANK AREAS

SAMPLE EXAM

1. A soil specimen has a void ratio of 0.6, moisture content of 10%, and $G = 2.67$. Determine the moist unit weight.
   a. 13.4 kN/m^3
   b. 15.6 kN/m^3
   c. 18 kN/m^3
   d. 20.1 kN/m^3

2. Calculate the seepage loss through the 0.5-m thick sandy silt layer below the earth dam shown (m^3/hr/m). *Given:* coefficient of permeability, $k = 3.66$ cm/hr.
   a. 0.0023
   b. 0.0037
   c. 0.0051
   d. 0.0093

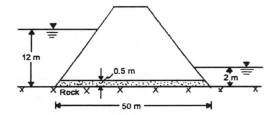

3. A normally consolidated clay layer having a thickness of 3 m is shown. A laboratory consolidation test on the same clay gave the following results:

p (kN/m^2)	e
75	0.661
150	0.557

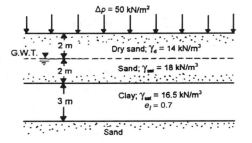

   The settlement of the clay layer due to a surcharge $\Delta p = 50$ kN/m^2 is closest to:
   a. 0.14 m
   b. 0.25 m
   c. 0.56 m
   d. 0.75 m

4. A silty clay layer of 6 m thickness is shown on the following page. It has a drained friction angle of 21° and a cohesion of 20 kN/m^2. What would be the drained shear strength at A (in kN/m^2)?
   a. 30
   b. 45
   c. 60
   d. 75

```
          ↑
       1.5 m       Dry sand; γ_d = 15 kN/m³
          ↓                                    ▼ G.W.T.
          ↑
        3 m        Sand; γ_sat = 18 kN/m³
          ↓
                 ↑
                2 m
             A ●  ⊥
        6 m        Silty clay; γ_sat = 18.75 kN/m³
          ↓
```

5. For a granular soil, the dry unit weight in the field is 16.35 kN/m³. The maximum and minimum dry unit weights of the same soil as determined in the laboratory are 17.6 kN/m³ and 14.46 kN/m³, respectively. Given $G = 2.66$, the relative density of compaction in the field is closest to
   a. 0.48
   b. 0.53
   c. 0.65
   d. 0.76

6. For a clayey soil, given:
   Plasticity index = 18
   Plastic limit = 15
   Specific gravity of soil solids, $G = 2.68$
   Determine the void ratio of the soil at liquid limit. *Note:* At liquid limit, the soil is saturated.
   a. 0.60
   b. 0.70
   c. 0.80
   d. 0.90

7. In a laboratory consolidation test on a 1-in-thick clay specimen, for a given pressure increment, the time for 50% consolidation is 2 min. The laboratory specimen is drained at the top and bottom. For a similar clay layer 20 ft thick in the field and for a similar pressure increment, the time for 50% consolidation will be about how many days? (*Note:* The clay layer in the field is drained at the top only.)
   a. 240
   b. 320
   c. 400
   d. 480

8. A sieve analysis for a soil sample is as follows:
   Passing 3 in sieve = 100%
   Passing No. 4 sieve = 95%
   Passing No. 200 sieve = 4%
   Also given:
   $D_{10} = 0.09$ mm
   $D_{30} = 0.185$ mm
   $D_{60} = 0.41$ mm

   Based on the Unified Soil Classification System, the soil can be classified as
   a. GW
   b. SP
   c. CH
   d. ML

9. A frictionless retaining wall is 10 ft high. It has a dry sand backfill. The angle of friction (ϕ) is 32°, and the dry unit weight is 115 lb/ft^3. The Rankine active force per unit length of the wall is approximately
   a. 1500 lb/ft
   b. 1750 lb/ft
   c. 2000 lb/ft
   d. 2250 lb/ft

10. The ratio of width b to depth y that maximizes the discharge through a fixed cross-sectional area in an open channel of rectangular section is
    a. $b/y = 1/3$
    b. $b/y = 1/2$
    c. $b/y = 1$
    d. $b/y = 2$

11. The depth of flow in a concrete-lined ($n = 0.012$) triangular channel, laid on a slope of 2 m per 3 km and carrying a discharge of 1.0 m^3/s, is closest to
    a. 0.85 m
    b. 0.91 m
    c. 0.97 m
    d. 1.03 m

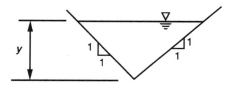

12. The hydraulic radius for a circular pipe of diameter D flowing half-full is
    a. $D/4$
    b. D
    c. $\pi D/2$
    d. πD

13. The troweled concrete channel shown conveys water at a depth of 1 m on a slope of 0.0004. The discharge is closest to
    a. 1.0 m^3/s
    b. 2.1 m^3/s
    c. 3.2 m^3/s
    d. 10.0 m^3/s

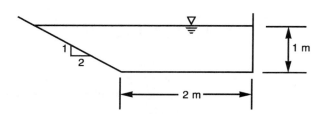

14. A 2-m-diameter pipe connecting two reservoirs has an interior surface of smooth masonry ($C = 120$) and is 5000 m long. The difference in reservoir water surface elevations is 6.0 m. According to the Hazen-Williams formula, the discharge is nearest to
   a. 1.7 m^3/s
   b. 2.7 m^3/s
   c. 5.5 m^3/s
   d. 8.5 m^3/s

15. A new cast iron pipe ($C = 130$) is to convey 4.5 m^3/s a distance of 6 km from one reservoir to another reservoir whose water surface is 7.5 m lower than the first reservoir. The most appropriate pipe diameter is
   a. 1.2 m
   b. 1.8 m
   c. 2.4 m
   d. 3.0 m

16. A concrete-lined open channel with a slope of 0.34% has a bottom width of 1.5 m and side slope of 2:1. This channel is to be designed to handle peak discharge of 15 m^3/s. Assume $n = 0.015$. The depth of flow in feet is nearest to:
   a. 1.0 m
   b. 1.2 m
   c. 1.4 m
   d. 1.6 m

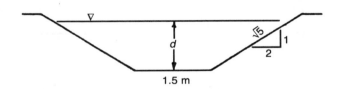

QUESTIONS 17-18

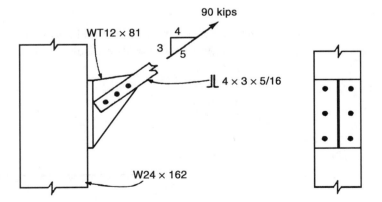

The relevant properties of one angle are $A_g = 2.09$ inches^2, hole diameter = $d_b + 0.125$ inches. The allowable stress may be increased by one-third when considering wind loads.

17. All bolts in the wind bracing connection shown in the figure are $\frac{7}{8}$-inch A325 slip-critical bolts in standard holes ($A_g = 2.09$). All structural sections are Grade A36 steel. The effective net area of the angle brace is most nearly:
   a. 2.0 square inches
   b. 2.5 square inches
   c. 3.0 square inches
   d. 3.5 square inches

18. The capacity of the angle brace based on its effective net area is most nearly: $F_u = 58$, $A_e = 3.02$
   a. 101 kips
   b. 106 kips
   c. 111 kips
   d. 116 kips

19. The W14 × 48 Grade A36 column shown below supports an axial load only. The relevant properties of the W14 × 48 are $A_s = 14.1$ inches^2, $E_s = 29{,}000$ ksi, $F_y = 36$ ksi, $r_x = 5.85$ inches, and $r_y = 1.91$ inches. The allowable axial load is most nearly:
   a. 230 kips
   b. 240 kips
   c. 250 kips
   d. 260 kips

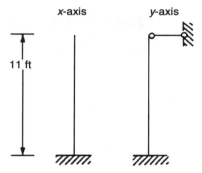

20. A 500-mm diameter, spirally reinforced column with 40 mm cover to the spiral is reinforced with 10 No. M30 bars. The concrete strength is 35 MPa and the tensile strength of the reinforcement is 400 MPa. The column may be classified as a short column.

    The column supports only axial load and the design axial load strength is given most nearly by:
   a. 4800 kN
   b. 5000 kN
   c. 5200 kN
   d. 5400 kN

21. For the column in Problem 20, the minimum permitted diameter of the spiral reinforcement is:
   a. $\frac{1}{4}$-inch
   b. $\frac{3}{8}$-inch
   c. $\frac{1}{2}$-inch
   d. $\frac{5}{8}$-inch

22. What is the required reinforcement in an 18 in × 18 in reinforced concrete column ($f'_c = 4$ ksi; $f_y = 60$ ksi) subjected to the following loads? $P_{DL} = 250$ kips, $P_{LL} = 450$ kips. Assume the eccentricity of the load is 4 in.
   a. 24.0 in^2
   b. 3.24 in^2
   c. 5.60 in^2
   d. 18.0 in^2

23. A 300-m-radius simple curve with a tangent deflection of 40° has been staked on the ground. It is desired to **sharpen** the curve so as to move the midpoint of the curve a distance of 3 m **radially**. The tangent deflection must remain fixed at 40° so the direction of the tangents is not changed. The radius (m) of the new curve is most nearly:
    a. 297
    b. 331
    c. 253
    d. 340

24. An existing vertical curve joins a +4% with a −4% grade. Assume $f = 0.4$, a perception-reaction time of 2.5 s, and that $S < L$. If the length of the curve is 75 m, the maximum safe speed (km/h) on the curve is most nearly:
    a. 50
    b. 40
    c. 60
    d. 45

QUESTIONS 25–26

Given the following traffic count data:

Time Interval	No. of Vehicles
5:00–5:15	1000
5:15–5:30	1100
5:30–5:45	1200
5:45–6:00	900

25. The peak hour factor is most nearly:
    a. 0.950
    b. 0.875
    c. 0.050
    d. 0.750

26. The peak rate of flow (veh/hr) is most nearly:
    a. 3600
    b. 4400
    c. 4800
    d. 3675

QUESTIONS 27–28

Given the following spiral and simple curve data: PI at station 1 + 086.271, $R = 300$ m, $\Delta = 16$ degrees, $\lambda_s = 52.083$ m, $p = 0.377$ m, $k = 26.035$ m.

27. Assuming a rate of increase of centripetal force (C) of 0.6 m/s^3, the recommended design speed (km/h) for the curve is most nearly:
    a. 100
    b. 60
    c. 120
    d. 75

28. The stationing of the TS for the curve is most nearly:
    a. 1 + 018
    b. 1 + 154
    c. 1 + 138
    d. 1 + 000

29. A section of highway has the following design requirements.
    Design speed = 80 km/hr
    Coefficient of friction = 0.35
    Perception-reaction time = 2.5 s
    Grade = +3%
    Driver eye height = 1070 mm
    Object height = 150 mm

    The stopping sight distance (m) for this highway is most nearly:
    a. 135
    b. 120
    c. 130
    d. 110

30. Direct and reverse zenith angle measurements at a station are recorded as follows:
    Direct reading   71° 40′ 12″
    Reverse reading   288° 19′ 40″

    The corrected zenith angle is
    a. 71° 40′ 16″
    b. 71° 40′ 08″
    c. 71° 40′ 20″
    d. 71° 40′ 04″

31. An EDM (electronic distance measurement) instrument is used to sight a prism located at station B. The distance is measured as 234.78 m, and the zenith angle is 73° 32′ 30″. If the height of instrument (HI) is 565.23 m above sea level, the elevation of the prism at B is (neglect curvature and refraction effects)
    a. 776.85 m above sea level
    b. 790.39 m above sea level
    c. 394.92 m above sea level
    d. 631.75 m above sea level

32. For the triangular traverse ABC, coordinates are given in the following table. What is the area of the triangular area ABC?

Station	Northing (m)	Easting (m)
A	+126.54	− 320.66
B	+467.19	+213.45
C	+621.56	+376.44

    a. 14,322 m^2
    b. 7161 m^2
    c. 13,464 m^2
    d. 26,928 m^2

QUESTIONS 33-34

For a particular site, the existing and the proposed ground profiles have been drawn to establish the following cross-sectional areas at stations spaced at 25-ft intervals.

Station	End Area (ft^2)
204 + 00.00	214.56
204 + 25.00	345.54
204 + 50.00	412.45
204 + 75.00	334.78
205 + 00.00	325.45
205 + 25.00	276.45
205 + 50.00	233.23
205 + 75.00	127.78

33. Using Simpson's rule, the volume of earthwork between stations 204 + 25.00 and 205 + 75.00 is
    a. 1684 yd^3
    b. 1500 yd^3
    c. 1722 yd^3
    d. 1903 yd^3

34. Two tangents intersecting at an angle of 35° 45′ are to be joined by a horizontal circular curve. If the degree of curve, based on a 100-ft chord, is 9° 30′, and the PI is at station 19 + 42, the stations of the PC and PT are
    a. 17 + 47.27, 21 + 24.01
    b. 17 + 47.27, 21 + 36.73
    c. 16 + 32.74, 23 + 18.74
    d. 16 + 32.74, 22 + 43.12

35. The minimum number of visible GPS satellites needed to determine the position of a point on the ground (latitude, longitude, and elevation) is
    a. 2
    b. 3
    c. 4
    d. 5

36. A distance is measured as 43.565 m using steel tape on a day when the ambient temperature is 27° C. The tape is held level between supports using a tension of 50 N. Use the following data:

    Standardization temperature = 20° C
    $\alpha = 11.6 \times 10^{-6}$/° C
    E = 205 GPa
    Tape cross-sectional area = 3 mm × 0.5 mm
    Tape weight per unit length = 0.12 N/m
    The tape length is standardized while supported on the ground with tension = 30 N

    The corrected distance is
    a. 43.591 m
    b. 43.552 m
    c. 43.539 m
    d. 43.579 m

QUESTIONS 37–38

A water has the following ionic composition:

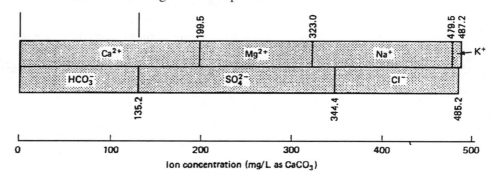

37. The total carbonate hardness as CaCO₃ is
    a. less than 100 mg/L
    b. greater than the noncarbonate hardness
    c. only associated with Ca^{2+}
    d. 244.4 mg/L

38. The $Ca(OH)_2$ needed to remove the carbonate hardness (CH) is
    a. 1 mol/mol CH
    b. 2.2 mg/mg CH
    c. 0.46 mg/mg CH
    d. dependent upon both the calcium and magnesium concentrations

39. A single source of BOD to a river causes a minimum DO downstream of 6 mg/L. The initial DO deficit is zero and DO saturation is 10 mg/L.
    a. Without aeration, the deficit will decrease.
    b. Doubling the BOD load would double the deficit.
    c. Doubling the BOD load would reduce the DO to 3 mg/L.
    d. Doubling the BOD load would move the point where the deficit occurs closer to the discharge.

40. A distilled water solution has 0.001 M H_2SO_4. The pH is
    a. 2.7
    b. 11.3
    c. neutral
    d. 3.0

41. Magnesium in an aqueous solution is in equilibrium with precipitated magnesium carbonate.
    1. The magnesium concentration is dependent upon the pH.
    2. The magnesium concentration is dependent upon the carbonate concentration.
    3. The magnesium concentration is dependent upon the bicarbonate concentration.
    4. The magnesium concentration is dependent upon alkalinity.

    Which of the above statements are correct?
    a. Statements 1 and 4 are correct.
    b. Statements 2 and 3 are correct.
    c. Statements 1 and 2 are correct.
    d. All of the statements are correct.

42. In using chlorine for disinfection
    a. concentration of HOCI is pH-dependent
    b. [HOCl] equals [OCl⁻] at pH 7
    c. effectiveness increases with time of contact
    d. total elimination of bacteria is possible

43. A 61-cm-diameter circular concrete sewer is set on a slope of 0.001. The flow when it is half-full is
    a. $\frac{1}{2}$ of flow at Q_{full}
    b. 0.08 m³/s
    c. 0.20 m³/s
    d. 0.28 m³/s

44. A rigid bar BCD is supported by a hinged connection at support B and by a cable AC as shown. The reaction force at support B is

    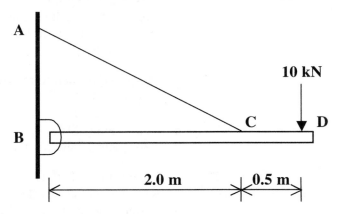

    a. 2.50 kN
    b. 14.20 kN
    c. 9.70 kN
    d. 14.42 kN

45. A simply supported beam AB is subjected to a uniformly distributed load and a concentrated load as shown. The midspan deflection is

    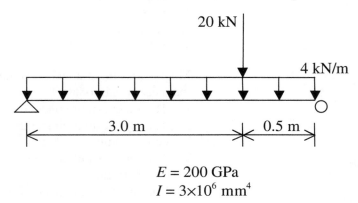

    $E = 200$ GPa
    $I = 3 \times 10^6$ mm⁴

    a. 16 mm
    b. 38 mm
    c. 22 mm
    d. 8 mm

46. A steel column (W21 × 93) has the bottom end supported rigidly so as to prevent both rotation and translation. The top of the column is free to translate and rotate in the plane of the diagram but pinned (rotation allowed, translation prevented) in the perpendicular direction. The length of the column is 32 ft. The critical buckling load as given by Euler's theory is
    a. 13.4 kips
    b. 36.7 kips
    c. 366 kips
    d. 1002 kips

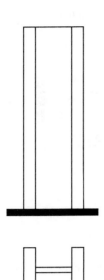

47. The three-hinged arch ABC carries a uniformly distributed load from a horizontal deck as shown. The value of the horizontal thrust at the supports is

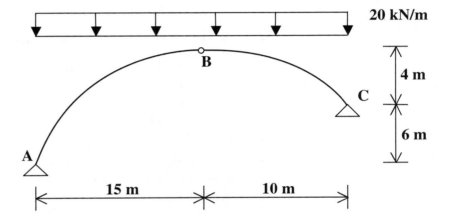

    a. 282 kN
    b. 197 kN
    c. 218 kN
    d. 252 kN

48. Given the truss shown, the vertical deflection at B is

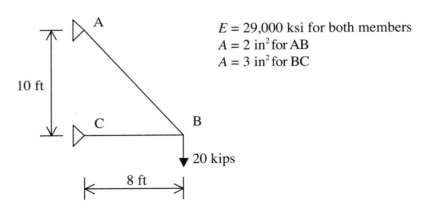

$E = 29{,}000$ ksi for both members
$A = 2$ in^2 for AB
$A = 3$ in^2 for BC

    a. 1.2 in
    b. 0.05 in
    c. 0.2 in
    d. 0.1 in

49. The maximum bending moment for the beam shown is

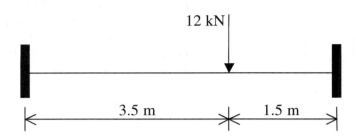

a. 12.60 kNm
b. 8.82 kNm
c. 9.41 kNm
d. 10.25 kNm

50. A client requires a new facility quickly to replace one destroyed by a hurricane. He is not certain exactly what it should include. The best type of contract to be used in this situation is
a. unit price
b. lump sum
c. cost plus fixed fee
d. surety

51. What is the height (ft) of a conical spoil pile, if 100 ft bank cu yd of common earth with an angle of repose of 32° and a 12% swell is deposited?
a. 9.6
b. 19.2
c. 10.4
d. 33.3

52. Determine the volume of fill for a whole station (100 ft) of cut with the end areas shown below.

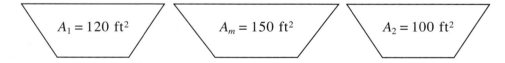

a. 820 yd^3
b. 456.8 yd^3
c. 506.2 yd^3
d. 569.4 yd^3

53. ABC Construction Company is the low bidder at $500,000 on a building project. The company has provided the typical bid bond. After award of the contract, the company decides to withdraw its bid. The second lowest bid was $520,000. How much is the surety company required to reimburse the client for the failure of ABC to sign?
a. $500,000
b. $50,000
c. $20,000
d. $5000

54. A contractor has bid a lump sum of $50,000 (direct costs) to install 200 linear feet of pipe, plus $5000 overhead and profit. The client wants the contractor to quote him a unit price for the work. What should the contractor charge per linear foot?
   a. $250
   b. $275
   c. $500
   d. $200

55. The best type of construction schedule to show a nontechnical client would most likely be
   a. A-O-A
   b. PERT
   c. A-O-N
   d. Gantt

56. A concrete mix uses a coarse aggregate that has a moisture content of 3.0% (dry basis). The quantity of wet coarse aggregate used in the mix is 55 kg. The moisture content of SSD aggregate is 0.7%. The specific gravity of SSD coarse aggregate is 2.70. How much additional water becomes available from the coarse aggregate?
   a. 0.375 kg
   b. 0.385 kg
   c. 1.650 kg
   d. 1.226 kg

57. A tensile test is performed on a steel sample as shown below. The sample is cut from a steel plate of 6.3 mm thickness. If the uniformly distributed tensile force is 15 kN, what is the elongation of the gage length?

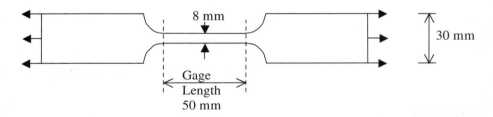

   a. 0.037 mm
   b. 0.073 mm
   c. 0.147 mm
   d. 0.295 mm

58. A concrete mix has the following characteristics:
   Weight of wet coarse aggregate = 125 lb
   Weight of available water in coarse aggregate = 4 lb
   Weight of wet fine aggregate = 95 lb
   Weight of available water in fine aggregate = 3 lb
   Weight of added water = 29 lb
   Volume of cement = 0.42 ft^3

   What is the water cement ratio? Assume specific gravity of cement is 3.10.
   a. 0.28
   b. 0.36
   c. 0.44
   d. 0.51

59. A rectangular concrete beam (width $b = 4$ in; depth $d = 6$ in) is tested in pure bending by using a two-point load as shown in the figure. The beam is unreinforced and the 28-day compression strength is $f_c' = 4500$ psi.

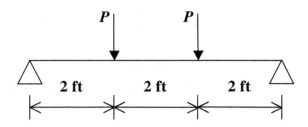

The ACI code gives the following empirical relationship for the rupture stress (tensile strength) of concrete: $f_r = 7.5\sqrt{f_c'}$ where f_r and f_c' are in psi. What is the maximum load P before the beam experiences tensile cracking?
a. 500 lb
b. 6000 lb
c. 250 lb
d. 3000 lb

60. An aggregate blend composed for an asphalt mix has the following composition:

Coarse aggregate (specific gravity = 2.75)	58%
Fine aggregate (specific gravity = 2.50)	42%

The bulk specific gravity of the aggregate blend is
a. 2.639
b. 2.645
c. 5.250
d. 2.625

End of Exam. Please check your work.

SOLUTIONS

1. **c.** Dry unit weight,

$$\gamma_d = \frac{G\gamma_w}{1+e} = \frac{(2.67)(9.81)}{1+0.6} = 16.37 \text{ kN/m}^3$$

Moist unit weight,

$$\gamma = \gamma_d(1+w) = (16.37)(1+0.1) = 18 \text{ kN/m}^3$$

2. **b.** $Q = kiA$

$$k = 3.66 \text{ cm/hr}; \quad i = \frac{12-2}{50} = 0.2; \quad A = (0.5)(1) = 0.5 \text{ m}^2$$

$$Q = \left(\frac{3.66}{100}\right)(0.2)(0.5) = 0.0037 \text{ m}^3\text{/hr/m}$$

3. **a.**

$$C_c = \frac{e_1 - e_2}{\log\left(\frac{p_2}{p_1}\right)} = \frac{0.661 - 0.577}{\log\left(\frac{150}{70}\right)} = 0.279$$

Effective overburden pressure at the middle of the clay layer,

$$p_i = (14 \times 2) + (18 - 9.81)(2) + \frac{3}{2}(16.5 - 9.18) = 54.42 \text{ kN/m}^2$$

$$\Delta H = \frac{C_c H}{1 + e_i} \log\left(\frac{p_i + \Delta p}{p_i}\right) = \frac{(0.275)(3)}{1 + 0.7} \log\left(\frac{54.42 + 50}{54.42}\right)$$
$$= 0.139 \text{ m} = 139 \text{ mm}$$

4. **b.**

$s = c + \sigma' \tan\phi$
$c = 20 \text{ kN/m}^2; \phi = 21; \sigma' = (15)1.5 + 3(18 - 9.81) + 2(18.75 - 9.81)$
$= 64.95 \text{ kN/m}^2$
$s = 20 + 64.95 \tan 21° = 44.9 \text{ kN/m}^2$

5. **c.**

$$D_d = \frac{\frac{1}{\gamma_{min}} - \frac{1}{\gamma_d}}{\frac{1}{\gamma_{min}} - \frac{1}{\gamma_{max}}} = \frac{\frac{1}{14.46} - \frac{1}{16.35}}{\frac{1}{14.46} - \frac{1}{17.6}} = 0.645 = 64.5\%$$

6. **c.** LL = PL + PI = 15 + 18 = 33
Degree of saturation,

$$S(\%) = \frac{wG}{e} \times 100$$

If $S = 100\%$,

$$e_{LL} = wG = \left(\frac{30}{100}\right)(2.68) = 0.804$$

7. **b.** Time factor, $T = \dfrac{c_v t}{H_{dr}^2}$

For 50% consolidation, $\dfrac{c_v t_{(lab)}}{H_{dr(lab)}^2} = \dfrac{c_v t_{(field)}}{H_{dr(field)}^2}$

$$t_{(lab)} = 2 \text{ min}$$
$$H_{dr(lab)} = \frac{1}{2} = 0.5 \text{ in}$$
$$H_{dr(field)} = 20 \text{ ft} = 240 \text{ in}$$

So
$$\frac{2 \text{ min}}{(0.5 \text{ in})^2} = \frac{t_{(field)}}{(240 \text{ in})^2}$$

$$t_{(field)} = \frac{2 \times (240)^2}{(0.5)^2} = 460{,}800 \text{ min} = 320 \text{ days}$$

8. **b.** Passing No. 4 sieve = 95%. So gravel fraction is 100 − 95 = 5%. Passing No. 200 sieve = 4%. So sand and gravel fraction (coarse fraction) = 100 − 4 = 96%.
Hence, sand fraction = 96 − 5 = 91%.
More than 50% of coarse fraction passing No. 4 sieve, so it is a sandy soil.

$$C_u = \frac{D_{60}}{10} = \frac{0.41}{0.09} = 4.56$$

$$C_c = \frac{D_{30}^2}{D_{60} \times D_{10}} = \frac{(0.185)^2}{(0.41)(0.09)} = 0.93$$

Since fines are 4% < 5%, $C_c = 0.93 < 1$, and $C_u = 4.56 < 6$, the soil is SP.

9. **b.** Rankine active force per unit length = $P_a = 0.5\gamma H^2 K_a$

$$K_a = \tan^2\left(45 - \frac{\phi}{2}\right) = \tan^2\left(45 - \frac{32}{2}\right) = 0.307$$

$$P_a = (0.5)(115)(10)^2(0.307) = 1765.25 \text{ lb/ft}$$

10. **d.** The Manning discharge formula is

$$Q = \frac{1}{n} A R^{\frac{2}{3}} S^{\frac{1}{2}}$$

For a fixed area A the discharge is maximized when R is maximized. With $A = by$ and the wetted perimeter $p = b + 2y$,

$$R = \frac{A}{P} = \frac{by}{b+2y} = \frac{A}{\frac{A}{y}+2y}$$

$$\frac{dR}{dy} = \frac{-A}{\left(\frac{A}{y}+2y\right)^2}\left(-\frac{A}{y^2}+2\right) = 0$$

or $A = 2y^2 = by$ so $b = 2y$, $b/y = 2$.

11. **c.** The discharge is

$$Q = \frac{1}{n} A R^{\frac{2}{3}} S^{\frac{1}{2}}$$

where

$$A = y^2$$
$$P = 2\sqrt{2}y$$
$$R = A/P = \frac{y}{(2\sqrt{2})}$$

Hence

$$1.0 = \frac{1}{0.012} y^2 \left(\frac{y}{2\sqrt{2}}\right)^{\frac{2}{3}} \left(\frac{2}{3000}\right)^{\frac{1}{2}}$$

or $y^{\frac{8}{3}} = 0.930$, which yields $y = (0.930)^{\frac{3}{8}} = 0.973$ m.

12. **a.** The hydraulic radius is $R = A/P$. Using $r = D/2$, for a half-full circular pipe one has $A = \pi r^2/2$, $P = \pi r$ so that

$$R = \frac{A}{P} = \frac{\frac{\pi r^2}{2}}{\pi r} = \frac{r}{2} = \frac{D}{4}$$

13. **c.** The discharge is

$$Q = \frac{1}{n} A R^{\frac{2}{3}} S^{\frac{1}{2}}$$

From the table of Manning n values, $n = 0.013$. Also,

$$A = 2 + \frac{1}{2}(1)(2) = 3.0 \text{ m}^2$$
$$P = 1 + 2 + \sqrt{5} = 5.24 \text{ m}$$

Therefore

$$Q = \frac{1}{0.013}(3.0)\left(\frac{3.0}{5.24}\right)^{\frac{2}{3}}(0.0004)^{\frac{1}{2}}$$

and

$$Q = 3.2 \text{ m}^3/\text{s}$$

14. **c.** For smooth masonry the Hazen-Williams coefficient is $C = 120$. The Hazen-Williams formula is

$$V = 0.849 C R^{0.63} S^{0.54}$$

Here

$$A = \frac{\pi}{4} D^2 = \frac{\pi}{4}(2)^2 = \pi \text{ m}^2$$
$$P = \pi D = 2\pi \text{ m}$$
$$R = \frac{A}{P} = 0.5 \text{ m}$$

Neglecting local losses, $S = h_L/L = 6/5000$. Hence

$$V = 0.849(120)(0.5)^{0.63}\left(\frac{6}{5000}\right)^{0.54} = 1.74 \text{ m/s}$$

and

$$Q = AV = \pi(1.74) = 5.47 \text{ m}^3/\text{s}$$

15. **b.** The Hazen-Williams equation may be written as

$$V = 0.849 C A R^{0.63} S^{0.54}$$

or as

$$AR^{0.63} = \frac{Q}{0.849 C S^{0.54}} = \frac{4.5}{0.849(130)\left[\frac{7.5}{6000}\right]^{0.54}} = 1.507$$

In this case

$$A = \frac{\pi}{4}D^2 \quad \text{and} \quad R = \frac{A}{P} = \frac{D}{4}$$

for a pipe flowing full. Thus

$$\frac{\pi}{4}D^2 \frac{D^{0.63}}{4^{0.63}} = 1.507$$

$$D^{2.63} = \frac{1.507(4^{1.63})}{\pi} = 4.59$$

and

$$D = (4.59)^{\frac{1}{2.63}} = (4.59)^{0.38} = 1.785 \text{ m}$$

16. b. Find the flow depth, d, in feet. Manning's formula

$$v = \frac{1}{n}R^{\frac{2}{3}}S^{\frac{1}{2}}$$

where
  v = velocity
  $R = A/P$
  A = channel cross-sectional area
  P = wetted perimeter

gives

$$Q = AV = \frac{1}{n}AR^{\frac{2}{3}}S^{\frac{1}{2}}$$

For this trapezoidal channel

$$A = (1.5 + 2d)d \quad \text{and} \quad P = 1.5 + 2\sqrt{5}d$$

Hence

$$15 = \frac{1}{0.015}(1.5+2d)d\left[\frac{(1.5+2d)d}{1.5+2\sqrt{5}d}\right]^{\frac{2}{3}}(0.0034)^{\frac{1}{2}}$$

Solve by trial for d; that is, choose a value of d, substitute into the equation and see whether it is satisfied. An example of the progress of this solution starting with $d = 1.0$ is shown in the table below.

d	A	P	R	$R^{\frac{2}{3}}$	$AR^{\frac{2}{3}} \stackrel{?}{=} 3.8587$
1.0	3.50	5.97	0.586	0.701	2.45
1.5	6.75	8.21	0.822	0.878	5.92
1.2	4.68	6.87	0.681	0.774	3.62
1.25	5.00	7.09	0.705	0.792	3.96
1.23	4.87	7.00	0.696	0.785	3.82

$$AR^{2/3} = 3.8587$$

where
$A = (1.5 + 2d)d$
$P = 1.5 + 2\sqrt{5}d$
$R = A/P$

We find $d = 1.23$ m, which gives $A = 4.87$ m^2.

17. **c.** From AISC Section B2, the net area of the brace is $A_n = A_g - 2td_h = 2 \times 2.09 - 2 \times 0.3125 (0.875 + 0.125) = 3.56$ square inches. From AISC Equation (B3-1), the effective net area of the brace, $A_e = UA_n = 0.85 \times 3.56 = 3.02$ square inches.

18. **d.** In accordance with AISC Section A5.2 a one-third increase in allowable stress is permitted for wind loading. The strength of the brace based on the effective net area, in accordance with AISC Section D1, is $P_t = 1.33 \times 0.5\, F_u A_e = 1.33 \times 0.5 \times 58 \times 3.02 = 116.48$ kips.

19. **c.** The relevant properties of the W14 × 48 are $A_s = 14.1$ inches^2, $r_x = 5.85$ inches, and $r_y = 1.91$ inches. The slenderness ratio about the x-axis is given by $Kl/r_x = 2.1 \times 11 \times 12/5.85 = 47.4$. The slenderness ratio about the y-axis is given by $Kl/r_y = 0.8 \times 11 \times 12/1.91 = 55.3$, which governs. $C_c = (2\pi^2 \times 29{,}000/36)^{0.5} = 126.1 > Kl/ry$. The allowable stress is $F_a = [1 - 55.3^2/(2 \times 126.1^2)]36/[5/3 + 3 \times 55/3/(8 \times 126.1) - 55.3^3/(8 \times 126.1^3)] = 17.88$ kips per square inch. The allowable axial load is $P = 14.1 \times 17.88 = 252.11$ kips.

20. **d.** For a spirally reinforced column, the design axial load strength is given by the equation as $\phi P_n = 0.85\, \phi[0.85\, f'_c(A_g - A_{st})\, f_y A_{st}] = 0.85 \times 0.75[0.85 \times 35(196{,}350 - 7000) + 4000 \times 7000] = 5376$ kN.

21. **b.** In accordance with ACI Section 7.10.4.2, the minimum diameter of the spiral reinforcement is $\frac{3}{8}$-inch.

22. **a.** $P_U = 1.2 \times 250 + 1.6 \times 450 = 1020$ kips
    $M_U = 1020 \times 4 = 4080$ kip-in
    Eccentricity greater than 0.1h (4 in > 0.1 × 18 in)
    Rectangular column must have tied reinforcement, not spiral. Thus, $\phi = 0.65$.
    Therefore, column must be designed for combination of P_U and M_U

    $$\frac{P_U}{\phi_c f'_c A_g} = \frac{1020}{0.65 \times 4 \times 324} = 1.21$$

    $$\frac{P_U e}{\phi_c f'_c A_g h} = \frac{1020 \times 4}{0.65 \times 4 \times 324 \times 18} = 0.27$$

    From the interaction diagram in the *FE Supplied-Reference Handbook*, $\rho_g = 7.4\%$.

    $$A_s = 0.074 \times 18 \times 18 = 23.98 \text{ in}^2$$

23. c. The relationship between the external ordinate (E) and the radius (R) is:

$$E = \left[\frac{R}{\cos\frac{\Delta}{2}} - R\right]$$

The external ordinate of the existing curve is:

$$E = \left[\frac{300}{\cos\frac{40}{2}} - 300\right] = 19.25 \text{ m}$$

The new curve is to be shifted radially a distance of 3 m. Therefore, the external ordinate of the new curve is 19.25 − 3 = 16.25 m. The new radius is

$$R = \frac{E}{\frac{1}{\cos\frac{\Delta}{2} - 1}} = \frac{16.25}{\frac{1}{\cos\frac{40}{2} - 1}} = 253.20 \text{ m}$$

24. a. Determine the stopping sight distance for the 75-m-crest vertical curve.

$$L = (AS^2)/404$$
$$75 = (8S^2)/404$$
$$S = 61.54 \text{ m}$$

Determine the maximum safe speed for this sight distance.

$$S = 0.278Vt + V^2/[254\,(f \pm g)]$$
$$61.54 = 0.278(V)(2.5) + V^2/[254\,(0.40 - 0.04)]$$

From the quadratic equation:

$$V = 49.69 \text{ km/h}$$

25. b.

$$\text{PHF} = V_h/(4V_{15})$$

where V_h = hourly volume (vph), V_{15} = maximum 15-minute rate of flow within the hour (veh).

$$\text{PHF} = (1000 + 1100 + 1200 + 900)/(4 \times 1200) = 0.875$$

26. c.

$$\text{Peak flow rate} = 4 \times 1200 = 4800 \text{ veh/hr}$$

27. d. The relationship between speed and length of spiral is:

$$\lambda_s = V^3/(46.7RC)$$
$$52.083 = V^3/[(46.7)(300)(0.6)]$$
$$V = 75.9 \text{ km/h}$$

28. a. Compute the spiral tangent.

$T_s = (R + p) \tan (\Delta/2) + k = (300 + 0.377) \tan (16/2) + 26.035 = 68.250$ m
TS station = PI station $- T_s = 1 + 086.271 - 0 + 068.250 = 1 + 018.021$

29. b. Stopping sight distance (S) is

$$S = 0.278Vt + V^2/[254(f \pm g)]$$
$$= 0.278(80)(2.5) + (80)^2/[254(0.35 + 0.03)] = 121.9 \text{ m}$$

30. a.
Direct reading	71° 40′ 12″
Reverse reading	288° 19′ 40″
Sum of both readings	359° 59′ 52″
360° minus sum	00° 00′ 08″
Error (half value)	00° 00′ 04″
Original reading	71° 40′ 12″
Corrected reading	71° 40′ 16″

31. d. Zenith angle = 73° 32′30″ (73.542°)
Elevation difference between sighting instrument and prism = $(234.78)\cos(73.542°) = + 66.52$ m
Elevation of prism = Elevation of sighting instrument + 66.52 = 565.23 + 66.52 = 631.75 m

32. c.

Station	Y (m)	X (m)	$X_{i-1} - X_{i+1}$	$Y_i (X_{i-1} - X_{i+1})$
A	+126.54	−320.66	162.99	20,624.75
B	+467.19	+213.45	−697.10	−325,678.15
C	+621.56	+376.44	534.11	331,981.41
				26,928.01

Area of traverse = $0.5 \times 26928.01 = 13{,}464 \text{ m}^2$

33. c.

i	y_i (ft^2)
0	345.54
1	412.45
2	334.78
3	325.45
4	276.45
5	233.23
6	127.78

$$V = \frac{\Delta}{3}\left[A_0 + A_n + 4\sum_{\text{odd}} A_i + 2\sum_{\text{even}} A_i\right]$$

$$V = \frac{25}{3}[345.54 + 127.78 + 4(412.45 + 325.45 + 233.23) + 2(334.78 + 276.45)]$$
$$= 46{,}502.5 \text{ ft}^3 = 1722.3 \text{ yd}^3$$

34. a. If the degree of curve is based on a 100-ft chord, the relationship is

$$100 = 2R\sin\frac{9.5°}{2} \Rightarrow R = \frac{100}{2\sin 4.75°} = 603.80 \text{ ft}$$

$$T = R\tan\frac{I}{2} = 603.80 \times \tan 17.875' = 194.73 \text{ ft} = 1 + 94.73 \text{ sta}$$

PI sta = PC sta + T = 19 + 42
Therefore, PC sta = (19 + 42) − (1 + 94.73) = 17 + 47.27
Length of curve L is the arc length corresponding to $R = 603.80$ ft and intersection angle = 35.75°.
$L = RI$ (I in radians)

$$L = 603.80 \times \frac{35.75°\pi}{180} = 376.74 \text{ ft} = 3 + 76.74 \text{ sta}$$

Therefore, PT sta = (17 + 47.27) + (3 + 76.74) = 21 + 24.01

35. c. If elevation of the point is also to be determined, the distance to four satellites is required. If elevation of the point is known, three satellites must be visible.

36. b. Temperature correction =

$$\alpha L \Delta T = 11.6 \times 10^{-6}/°C \times 43.56 \text{ m} \times 7°C = +0.00354 \text{ m}$$

Tension correction =

$$\frac{(P - P_0)L}{AE} = \frac{(50 - 30) \text{ N} \times 43.56 \text{ m}}{(0.003 \times 0.0005) \text{ m}^2 \times 205 \times 10^9 \text{ N/m}^2} = +0.00283 \text{ m}$$

$$\text{Sag correction} = \frac{w^2 L^3}{24 P^2} = \frac{(0.12 \text{ N/m})^2 \times (43.56 \text{ m})^3}{24 \times (50 \text{ N})^2} = -0.01984 \text{ m}$$

Corrected length = 43.565 + 0.00354 + 0.00283 − 0.01984
= 43.565 − 0.01347 = 43.552 m

37. c.

38. a.

39. b.

40. a.

41. d. All of the statements are correct.

42. a.

43. b.
$$Q_f = 0.2 \text{ m}^3/\text{s}$$
$$Q/Q_f = 0.4$$
$$Q = 0.08 \text{ m}^3/\text{s}$$

44. d. The free body diagram for BCD is shown. Note that a common mistake could be assuming that cable AC forms a 3-4-5 triangle when it doesn't. The vertical and horizontal components of T are $0.66T$ and $0.75T$, respectively.

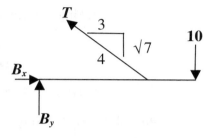

$$\sum M_B = 0.66T \times 2 - 10 \times 2.5 = 0 \Rightarrow T = 18.94$$

$$\sum F_x = B_x - 0.75 \times 18.94 = 0 \Rightarrow B_x = 14.20$$

$$\sum F_y = B_y + 0.66 \times 18.94 - 10 = 0 \Rightarrow B_y = -2.50$$

$$R_B = \sqrt{B_x^2 + B_y^2} = \sqrt{14.2^2 + 2.5^2} = 14.42$$

45. b. $EI = \left(2 \times 10^8 \dfrac{\text{kN}}{\text{m}^2}\right) \times (3 \times 10^{-6} \text{m}^4) = 600 \text{ kNm}^2$

Deflection at midspan ($x = 2$ m) due to point load is given by

$$\delta = \frac{Pb}{6LEI}[-x^3 + (L^2 - b^2)x] = \frac{20 \times 0.5}{6 \times 4 \times 600}[-2^3 + (4^2 - 0.5^2) \times 2] = 0.016$$

Deflection at midspan ($x = 2$ m) due to distributed load is given by

$$\delta = \frac{5wL^4}{384EI} = \frac{5 \times 4 \times 4^4}{384 \times 600} = 0.022$$

Total midspan deflection = 0.038 m = 38 mm

46. c. Buckling about the strong (x) axis:

$$K = 2.0; \; L = 32 \text{ ft}; \; r_x = 8.70 \text{ in}; \; KL/r = 88.3$$

Buckling about the weak (y) axis:

$$K = 0.7; \; L = 32 \text{ ft}; \; r_y = 1.84 \text{ in}; \; KL/r = 146.1$$

$$P_E = \frac{\pi^2 EA}{\left(\frac{KL}{r}\right)^2} = \frac{\pi^2 \times 29{,}000 \times 27.3}{146.1^2} = 366 \text{ kips}$$

47. b. There are four external reactions—A_x, A_y, C_x, and C_y—and two internal hinge forces—B_x and B_y—to be determined. Recognize that A_x and C_x are equal and opposite. Solving for either A_x or C_x is sufficient.

Taking moments about A (entire structure):

$$\sum M_A = 500 \times 12.5 + 4C_x - 25C_y = 0$$

Taking moments about B (right half of structure):

$$\sum M_{B,right} = 200 \times 5 - 6C_x - 10C_y = 0$$

Solving these equations, we get $C_x = -197.4$ kips.

48. d. Step 1: Using method of joints at joint B:

$$F_{AB} = +25.6 \text{ k (T)}$$
$$F_{BC} = -16.0 \text{ k (C)}$$

Step 2: Calculate the member forces due to virtual load (note: in this case, the real load and the virtual load look similar, so we can use scaling):

$$f_{AB} = +1.28 \text{ (T)}$$
$$f_{BC} = -0.80 \text{ (C)}$$

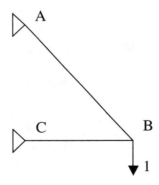

Step 3:

$$\Delta = \sum \left(\frac{FfL}{AE}\right) = \frac{25.6 \times 1.28 \times (12.81 \times 12)}{2 \times 29{,}000} + \frac{-16 \times -0.8 \times (8 \times 12)}{3 \times 29{,}000} = 0.10 \text{ in}$$

49. b. Using the fixed end moment load cases in the *FE Supplied-Reference Handbook* with the following data:

$$P = 12 \text{ kN};\ a = 3.5 \text{ m};\ b = 1.5 \text{ m};\ L = 5 \text{ m}$$

$$FEM_{AB} = \frac{Pab^2}{L^2} = \frac{12 \times 3.5 \times 1.5^2}{5^2} = 3.78 \text{ kNm}$$

$$FEM_{BA} = \frac{Pa^2b}{L^2} = \frac{12 \times 3.5^2 \times 1.5}{5^2} = 8.82 \text{ kNm}$$

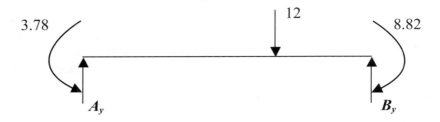

Taking moments about A,

$$\sum M_A = +3.78 - 12 \times 3.5 - 8.82 + 5B_y = 0 \Rightarrow B_y = 9.408 \text{ and } A_y = 2.592$$

The shear diagram and bending moment diagrams are:

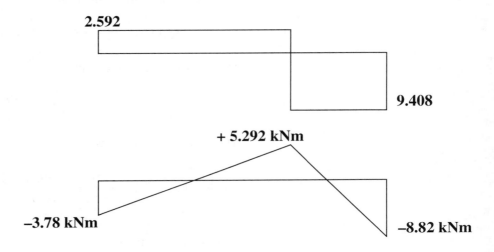

50. **c.** Cost plus fixed fee. Choices a and b require a detailed design; choice d is a bonding company.

51. **c.** Loose cu ft = 100 cu yd × 1.12 × 27 cu ft/cu yd = 3024 cu ft
    Base diameter (B) = $(7.64 \times 3024 / \tan 32°)^{1/3}$ = 33.3 ft
    Height = (33.3/2) (tan 32°) = 10.4 ft

52. **d.** This problem can be solved using the prismoidal method. Volume = (1 cu yd / 27 cu ft) 100 ft [120 sq ft + (4 × 150 sq ft) + 100 sq ft] / 6 = 506.2 cu yd

53. **c.** The client can only recover the loss incurred by the default, which is the difference between the low bid and the second lowest bid: $20,000.

54. **b.** Total cost = $50,000 + $5000 = $55,000
    Unit cost = $55,000 / 200 = $275.00/ft Note that the unit price must include the contractor's profit/markup.

55. d. The methods in the first three answer choices show detail, but a non-technical client may not understand them. The Gantt chart is the simplest to understand.

56. d. Weight of wet coarse aggregate = 55 kg (this represents 103% of the dry aggregate)
Weight of dry coarse aggregate = 55/1.03 = 53.4 kg
Weight of water in SSD aggregate (MC = 0.7%) = 0.007 × 53.4 = 0.374 kg (surface adsorbed water)
Therefore, available water = 55.0 − 53.4 − 0.374 = 1.226 kg

57. b.

Axial elongation

$$\Delta = \frac{PL}{AE} = \frac{15{,}000 \times 0.05}{(0.0063 \times 0.008) \times 205 \times 10^9} = 7.26 \times 10^{-5} \text{ m} = 0.0726 \text{ mm}$$

58. c. Total weight of mixing water in mix = 29 + 4 + 3 = 36 lb
Weight of cement = 0.42 × 3.1 × 62.4 = 81.24 lb
Water cement ratio = 36/81.24 = 0.44

59. a. For f'_c = 4500 psi, $f_r = 7.5\sqrt{f'_c} = 7.5\sqrt{4500} = 503.1$ psi

Maximum bending moment

$$M_{max} = \frac{PL}{3}$$

Maximum bending stress

$$\sigma_{max} = \frac{M_{max}}{S} = \frac{M_{max}}{\frac{bh^2}{6}} = \frac{PL/3}{bh^2/6} = \frac{2PL}{bh^2} \leq f_r \Rightarrow P = \frac{bh^2 f_r}{2L}$$

Therefore, the maximum load

$$P = \frac{bh^2 f_r}{2L} = \frac{4 \times 6^2 \times 503.1}{2 \times 72} = 503.1 \text{ lbs}$$

60. a.

$$G_{sb} = \frac{1}{\frac{P_c}{G_c} + \frac{P_f}{G_f}} = \frac{1}{\frac{0.58}{2.75} + \frac{0.42}{2.50}} = 2.639$$

INDEX

A
Actions, 11
Activated carbon, water treatment, 94
Alkalinity, 88
Allowable bending stress, 27
Atterberg limits, 3
Azimuth, 106–7

B
Bars, 15
Beams
 reinforced concrete, 35–43
 steel, elastic design of, 27–31
Benchmark, 106
Bending stresses, 27–31
Biochemical oxygen demand (BOD), 49, 89
Breakpoint chlorination, 57

C
Carbonate alkalinity, 88
Chemical oxygen demand (COD), 49
Chlorination
 wastewater, 57
 water, 93
Clayey soil, consistency of, 3
Columns
reinforced concrete, 43–44
 slenderness ratio of, 32
Combined chlorine residual, 57
Compass rule, 109
Compression members, structural steel design, 31–34
Consistency, clayey soils, 3
Consolidation settlement, 5
Construction management, 121–34
 contracts, 124–25
 estimating earthwork, 125–29
 procurement methods, 122–23
 productivity of equipment, 129
 project scheduling, 129–34
Contracts
 law, 124–25
 types of, 124
Coordinate systems
 global positioning system, 104
 state plane system, 103–4

D
Deflection, in reinforced concrete beams, 40–41
Departure, 107
Designing for strength, principles of, 35
Determinate structural analysis, 11
defined, 12
Differential leveling, 105–6
Directional design hourly volume (DDHV), 73
Discrete sedimentation, 51–52
Disinfection, 56–57
Dissolved oxygen concentration, water quality, 89–90
Domestic wastewater flows, 47
Dynamics, 12

E
Earthwork, 74–76, 125–29
Eastings, 108
Equilibrium relationships, water quality, 86
Examination, sample of, 139–58
Exchange capacity, 93

F
Factored moment, 37
Factored shear force, 41
Federal Water Pollution Control Act Amendments of 1972, 49
Filtration, 92
Flexure, reinforced concrete beams, 35–40
Flocculent sedimentation, 52
Flow nets, soil, 4–5
Fluoridation, water, 94
Forces, 11
Frames, 17–18
Free body, 13
Free body diagram (FBD), 13–15

G
Groundwater, 83, 91

H
Hazen-Williams equation, 25
Henry's Law, 86, 89
Heterogeneous processes, wastewater treatment, 56
Highway curves
 simple (circular) horizontal, 63–65
 spiral, 68
 vertical, 65–70
Homogeneous processes, wastewater treatment, 54–56
Hydraulics
 Hazen-Williams equation, 25
 Manning equation, 23–25
 of sewers, 47–48

I
Ion balances, water quality, 85
Ion exchange process, for water softening, 93

J
Joints, 15

L
Latitude, 107
Linear cut/fill, 128–29
Liquid limit, 3

M
Manning equation, 23–25, 47
Method of joints class, 15
Method of sections class, 15
Mixing, water quality, 90
Mohr-Coulomb failure criteria, 6
Moments, 11
Motion
 Newton's laws of, 11–12
 types of, 11

N
Newton's laws of motion, 11–12
Nitrogenous oxygen demand, 49
Normally consolidated soil, 5
Northings, 108

O
Overconsolidated soil, 5
Oxidation-reduction reactions, water quality, 85
Oxygen deficit, 89
Oxygen demand, wastewater, 49
Oxygen sag model, 90

P
Particle size distribution, soil, 1
Peak-hour factor (PHF), 73
Permeability, soil, 4
Pit excavation, 128
Plastic limit, 3
Plasticity index, 3
Present effective overburden pressure, 5
Prismoidal formula, 75, 128–29
Process analysis, wastewater treatment
 reaction rates, 50
 reactor types, 50–51
 types of reactions, 50

Project scheduling, 129–34
  critical path method, 129–32
  Gantt/bar chart, 129
  PERT, 133–34

R

Reinforced concrete beams
  deflection, 40–41
  flexure of, 35–40
  shear in, 41–43
Reinforced concrete design
  beams, 35–43
  columns, 43–44
  strength design principles for, 35–44
Relative density, soil, 3
Reoxygenation, 90

S

Sample examination, 139–58
Sedimentation
  wastewater treatment, 51–54
  water treatment, 91–92
Sewers, hydraulics of, 47–48
Shear strength, soil, 6
Shear stress, steel beams, 30–31
Shear, in reinforced concrete beams, 41–43
Shrinkage index, 3
Shrinkage limit, 3
Sight distance
  defined, 70
  on simple horizontal curves, 70–71
  on vertical curves, 71–73
Simple horizontal highway curves, 63–65
  sight distance on, 70–71
Simpson's rule, 113
Sludge, in water treatment, 94
Soil
  clayey, consistency of, 3
  consolidation settlement, 5
  flow nets, 4–5
  particle size distribution, 1
  permeability, 4
  relative density, 3
  shear strength, 6
  weight-volume relationships, 2, 125–26
Solubility relationships, water quality, 88–89
Space mean speed, 73
Speed, 73–74
Spiral highway curves, 68
Spoil banks/piles, 126–27
Static equilibrium, 11
Statics, 11
  central principle of, 12
Steel beams
  bending stresses, 27–30
  elastic design of, 27–31
  shear stress, 30–31
Stokes' law, 51

Stopping sight distance, 70
Structural analysis process
  frames, 17–18
  free body diagram (FBD), 13–15
  Newton's laws and, 12
  steps in, 12
  trusses, 15–17
Structural steel design
  compression members, 31–34
  steel beams, elastic design of, 27–31
  tensile stress, 34–35
Surface water, 83, 91
Surveying, 101–14
  angles and distances, 106–7
  area computations, 111–14
  basic trigonometry, 103
  coordinate systems, 103–4
  chaining, 104–5
  leveling, 105–6
  stationing, 104
  terms, 102–3
  traverse closure, 109–11
  types of surveys, 103

T

Tensile stress, structural steel design, 34–35
Theoretical oxygen demand (THOD), 49
Time mean speed, 73
Torques, 11
Total alkalinity, 88
Total suspended solids (TSS), 49
Traffic characteristics
  density, 74
  speed, 73–74
  traffic volume, 73
Traffic density, 74
Traffic volume, 73
Transit rule, 110–11
Transportation engineering
  earthwork, 74–76
  highway curves, 63–70
  sight distance, 70–73
  traffic characteristics, 73–74
Trapezoidal rule, 113
Trusses
  classes for analysis for, 15
  criteria for, 15
Twists, 11

U

Unconfined compression strength, 6

V

Vector, characteristics of, 11
Vertical highway curves, 65–70
  sight distance on, 71–73
Volume relationships, soil, 2

W

Wastewater
  characteristics of, 48–49
  oxygen demand, 49
  typical composition of, 48
Wastewater flows, 47
Wastewater physical processes, 51–54
Wastewater treatment
  biological processes, 54–56
  disinfection, 56–57
  federal mandates for, 49
  process analysis, 50–51
  sewer design, 47–48
  wastewater flows, 47
  wastewater, characteristics of, 48–49
Water distribution systems
  distribution network, 84
  pumping requirements, 85
  transmission line, 84
  water flow rates, 84
  water source, 83
  water storage, 84–85
Water quality
  alkalinity, 88
  common elements and radicals influencing, 86
  determinants of, 83
  dissolved oxygen relationships and, 89–90
  general chemistry concepts related to, 85–89
  ion balances in, 85
  solubility relationships, 88–89
Water softening, 92–93
  ion exchange process, 93
  with lime/soda ash, 93
Water source, 83
Water storage, 84–85
Water treatment
  activated carbon in, 94
  chlorination, 93–94
  common half-reactions for, 88
  common inorganic chemicals for, 87
  filtration, 92
  fluoridation, 94
  purpose of, 91
  sedimentation, 91–92
  sludge treatment, 94
  softening, 92–93
  types of, 91
Weight relationships, soil, 2

Z

Zone settling, 52–54